Concrete Materials 2018

Transportation Research Record: Journal of the Transportation Research Board

Volume 2672, Issue 27

ISBN: 978-1-5108-7831-0

Printed from e-media with permission by:

Curran Associates, Inc.
57 Morehouse Lane
Red Hook, NY 12571

Some format issues inherent in the e-media version may also appear in this print version.

Published by Sage Publications USA

ISBN (Print) 978-1-5108-7831-0
ISBN 2018 Printed Set (All Issues) 978-1-5108-7735-1

Printed by Curran Associates, Inc. (2019)

Additional copies of this publication are available from:

Curran Associates, Inc.
57 Morehouse Lane
Red Hook, NY 12571 USA
Phone: 845-758-0400
Fax: 845-758-2633
Email: curran@proceedings.com
Web: www.proceedings.com

Volume 2672 Number 27 December 2018

TRR
JOURNAL OF THE TRANSPORTATION RESEARCH BOARD

Contents

Concrete Materials

Articles

TRB

TRANSPORTATION RESEARCH BOARD

CORE SPONSORS

Transportation Departments of the 50 States and the District of Columbia

Federal Government

U.S. Department of Transportation
- Federal Aviation Administration (FAA)
- Federal Highway Administration
- Federal Motor Carrier Safety Administration (FMCSA)
- Federal Railroad Administration (FRA)
- Federal Transit Administration
- National Highway Traffic Safety Administration (NHTSA)
- Office of the Assistant Secretary of Transportation for Research and Technology (OST-R)

U.S. Air Force Civil Engineer Center (AFCEC)
U.S. Coast Guard (USCG)
U.S. Department of Energy
U.S. Department of the Interior (DOI)
U.S. Environmental Protection Agency (EPA)

Nongovernmental Organizations

American Public Transportation Association (APTA)
Association of American Railroads (AAR)
California Air Resources Board

The National Academies of
SCIENCES · ENGINEERING · MEDICINE

The **National Academy of Sciences** was established in 1863 by an Act of Congress, signed by President Lincoln, as a private, nongovernmental institution to advise the nation on issues related to science and technology. Members are elected by their peers for outstanding contributions to research. Dr. Marcia McNutt is president.

The **National Academy of Engineering** was established in 1964 under the charter of the National Academy of Sciences to bring the practices of engineering to advising the nation. Members are elected by their peers for extraordinary contributions to engineering. Dr. C. D. Mote, Jr., is president.

The **National Academy of Medicine** (formerly the Institute of Medicine) was established in 1970 under the charter of the National Academy of Sciences to advise the nation on medical and health issues. Members are elected by their peers for distinguished contributions to medicine and health. Dr. Victor J. Dzau is president.

The three Academies work together as the National Academies of Sciences, Engineering, and Medicine to provide independent, objective analysis and advice to the nation and conduct other activities to solve complex problems and inform public policy decisions. The Academies also encourage education and research, recognize outstanding contributions to knowledge, and increase public understanding in matters of science, engineering, and medicine.

Learn more about the National Academies of Sciences, Engineering, and Medicine at **www.national-academies.org.**

The **Transportation Research Board** is one of seven major programs of the National Academies of Sciences, Engineering, and Medicine. The mission of the Transportation Research Board is to increase the benefits that transportation contributes to society by providing leadership in transportation innovation and progress through research and information exchange, conducted within a setting that is objective, interdisciplinary, and multimodal. The Board's varied activities annually engage about 7,000 engineers, scientists, and other transportation researchers and practitioners from the public and private sectors and academia, all of whom contribute their expertise in the public interest. The program is supported by state transportation departments, federal agencies including the component administrations of the U.S. Department of Transportation, and other organizations and individuals interested in the development of transportation.

Learn more about the Transportation Research Board at **www.TRB.org**.

TRR
JOURNAL OF THE TRANSPORTATION RESEARCH BOARD

Article

Transportation Research Record
2018, Vol. 2672(27) 1–12
© National Academy of Sciences:
Transportation Research Board 2018
Article reuse guidelines:
sagepub.com/journals-permissions
DOI: 10.1177/0361198118798986
journals.sagepub.com/home/trr
SAGE

Flexural Performance Evaluation of Fiber-Reinforced Concrete Incorporating Multiple Macro-Synthetic Fibers

Michael Dopko[1], Meysam Najimi[1], Behrouz Shafei[1], Xuhao Wang[2], Peter Taylor[2], and Brent M. Phares[1]

Abstract

Fiber-reinforced concrete (FRC) is a promising construction material mainly because of the crack-controlling mechanisms that discrete fibers can impart to inherently brittle concrete. Macrofibers, in particular, have been proven effective for providing post-crack ductility and toughness, while synthetic fibers are a promising solution to avoid corrosion-related durability issues. To assess the performance enhancement provided by macro-synthetic concrete fibers, this study performs flexural tests on FRC beams containing three different types of macro-synthetic fibers. The selected fibers include polypropylene (PP), polyvinyl alcohol (PVA), and alkali-resistant glass (ARG) macrofibers mixed at volume fractions of 0.5%, 1.0%, and 1.5%. Static and dynamic fresh properties are monitored using the vibrating Kelly ball (VKelly) test. Beam specimens are then placed under a third point bending configuration, as per ASTM C1609 Standard, to measure load versus mid-span deflection. Strength and toughness parameters are derived from the load–deflection data to assess the flexural performance of the FRC composite systems under consideration. The parameters of interest include first peak strength (pre-crack flexural strength) and post-crack residual strength and toughness provided by fiber addition. Of the mixtures tested, ARG fiber mixtures show the highest residual strength and toughness values, followed by PP and PVA fiber mixtures. ARG fibers produce the most workable mixtures at all fiber volumes, while PVA fibers show a tendency to encounter dispersion issues at higher volume doses. The outcome of this study is expected to facilitate the selection of fibers by giving insight into their relative contribution to fresh and hardened flexural properties of FRC.

Formation and propagation of cracks in concrete is a complex process, which starts with the development of micro cracks under tensile stresses that exceed the concrete's tensile strength. Micro cracks then grow together and form macro cracks, which can propagate under tensile stresses until the crack becomes unstable and fracture occurs. To control the cracks and avoid a brittle failure mode, continuous reinforcement in the form of rebar or welded wire mesh is conventionally placed in the concrete where a tensile stress is expected. To further help avoid brittle failure and control crack formation and propagation in concrete, short fibers can also be dispersed randomly throughout the concrete matrix during the mixing process. In certain applications, such fibers not only reduce the required thickness, but potentially eliminate the need for continuous reinforcement, especially in concrete slabs.

The size, type, and other physical characteristics of fibers used in concrete are known to greatly affect the mixture properties in both fresh and hardened states. One common way of categorizing concrete fibers is by their size. In general, for conventional fiber-reinforced concrete (FRC), small "micro" fibers are used to prevent early-age cracking and increase the pre-crack performance of the matrix by arresting micro cracks, as they form. On the other hand, large "macro" fibers are included to maintain load carrying capacity and toughness after a crack is formed, as well as to control the propagation of cracks that have grown past the micro stage (*1*). Another way of categorizing concrete fibers is by the material they are made from. Steel fibers have shown many benefits to the mechanical properties of FRC (*2*),

[1]Institute for Transportation (InTrans), Department of Civil, Construction, and Environmental Engineering, Iowa State University, Ames, IA
[2]National Concrete Pavement Technology Center, Institute for Transportation (InTrans), Department of Civil Construction and Environmental Engineering, Iowa State University, Ames, IA

Corresponding Author:
Address correspondence to Behrouz Shafei: shafei@iastate.edu

but they face corrosion issues in the presence of harsh environmental conditions that civil structures are frequently exposed to. Natural fibers in concrete have received significant attention lately; however, most of them face certain limitations, such as high water absorption and lack of long-term stability in the highly alkaline concrete environment (*3*). The apparent limitations of steel and natural fibers in concrete have resulted in a growing interest in synthetic fibers that can avoid corrosion and strength degradation, even after a long period of time. Carbon, aramid, polyethylene, nylon, polypropylene, polyolefin, and polyvinyl alcohol are among synthetic fibers that have been explored for use as concrete fibers for a variety of applications (*4–9*).

In the situations that the hardened concrete is expected to exhibit high post-crack (residual) strength and toughness, as well as resistance against deterioration due to chemical instability or corrosion, macro-synthetic fibers are a sensible addition to the mixture. There is an abundance of literature that highlights the flexural and tensile performance of a single synthetic fiber type (*10–12*) or multiple fiber types mixed together (*13*, *14*). Although some studies have compared different macro-synthetic fiber types (e.g., *15*, *16*), there is a need for an in-depth study pertaining to the comparison of workability, flexural strength, and toughness provided by different macro-synthetic fibers that have potential to provide high residual capacity in concrete. To address this research gap, the current study investigates three types of synthetic macrofibers dispersed in a consistent concrete matrix. Static and dynamic fresh properties are monitored using the vibrating Kelly ball (VKelly) test, and hardened properties are evaluated using the flexural testing method in accordance with ASTM C1609 Standard (*17*). The fibers of choice include polypropylene (PP), polyvinyl alcohol (PVA), and alkali-resistant glass (ARG) macrofibers with reasonably similar aspect ratios. These three fiber types are selected based on their availability in the current concrete market and potential to provide high post-crack strength and toughness. The findings of this study provide insight into the capabilities and limitations of different types of market-available synthetic macrofibers, which further advance the state of the FRC knowledge and practice through a comparative analysis.

Materials

Fibers

Three types of macro-synthetic concrete fibers with comparable aspect ratios (i.e., PP, PVA, and ARG) are chosen to assess their flexural reinforcing potential as per ASTM C1609 Standard. PP fibers are one of the most common types of synthetic concrete fibers because of their relatively low cost, chemical stability in alkaline environment, and availability. Limitations of PP include low modulus of elasticity and poor bond with the concrete matrix. Bond properties of PP fibers can be enhanced by manufacturing the fibers with a twisted shape or texturing the fiber surface. Fibrillated micro PP fibers are a common concrete micro fiber mostly used to control cracking induced by plastic shrinkage in concrete applications with a large exposed surface (*18*). The macro PP fibers used in this study are a blend composed mostly of twisted bundle monofilament macrofibers, as well as a low volume of fibrillated strands.

PVA fibers are gaining popularity in the concrete market for several applications. PVA was initially developed to replace asbestos in making fiber cement using the Hatschek process (*19*). This type of fiber is very stable in the chemistry of the concrete environment, has a relatively high modulus of elasticity, and is reported to have the unique ability to chemically bond with the concrete matrix (*20*). Because of the superior bond properties of PVA, the fibers used in concrete are typically monofilament with no surface deformations. Micro PVA fibers are a key ingredient in Engineered Cementitious Composite (ECC), which utilizes a high volume of PVA fibers and fly ash to form a hardened mortar that is capable of up to 5% tensile strain. This is because of multiple, closely spaced micro cracks, which provide a pseudo strain hardening behavior under tensile strains (*21*). The PVA fibers used in this study are monofilament, straight, and have no surface deformations.

ARG fibers are made from extruding melted silicates with an optimum level of Zirconia into filaments and forming strands of the desired size to be cut to the desired length (*22*). Chopped ARG fibers made from this process tend to have a small diameter and high aspect ratio, categorizing them as a microfiber. They are also known to experience strength degradation in the alkaline environment of the concrete, much like basalt fibers (*23*), limiting their applications to early-age plastic shrinkage crack control. More recently, an ARG macrofiber has been developed by impregnating multiple ARG fiber filaments with a highly alkaline-resistant resin. The resin coating also helps improve the fibers' ability to bond with the concrete matrix. The fibers have a subtle twisted shape along their length to increase mechanical bond to the concrete matrix. This type of fiber has shown high post-crack load bearing capacity and toughness in FRC. Such fibers are also reported to disperse well in concrete because of their density being similar to that of the concrete matrix (*24*). The ARG fibers used in this study are composite monofilament, made with resin-impregnated strands and have a subtle twist along their length. Table

Table 1. Properties of Three Fiber Types Investigated in the Current Study

Fiber type	PP	PVA	ARG
Diameter (in. [mm])	0.013 (0.34)	0.008 (0.20)	0.026 (0.65)
Length (in. [mm])	1.50 (38)	0.71 (18)	1.69 (43)
Aspect ratio	112	90	66
Tensile strength (ksi [GPa])	90 (0.62)	145 (1.00)	157 (1.08)
Modulus of elasticity (ksi [GPa])	690 (4.7)	3920 (27)	6382 (44)
Specific gravity	0.9	1.3	2.1

1 lists the main properties of the three fiber types used in the current study. Because 3/8-in. maximum aggregate size is used, fiber length is set to be at least 3/4 in. to increase the fiber's reinforcing potential around the largest aggregate particles.

Concrete Matrix Composition

The base concrete mixture proportions are chosen to favor workability and reinforcing potential of the selected fibers. Type I/II Portland cement is used with 30% cement replacement with Class F fly ash to improve workability. The mixture uses 3/8-in. (9.5 mm) maximum size crushed limestone for coarse aggregate and river sand for fine aggregate. The 3/8-in. maximum size aggregate is chosen to maximize the volume within the matrix in such a way that fibers can exist without their position and orientation being restricted. This also improves the workability of FRC mixtures, as limiting the maximum aggregate size increases the volume space available for the fibers when the mixture is in the fluid state (*25*).

The water to cementitious materials ratio is held constant at 0.38 for all the mixtures. Because each of the three fibers tested has a different surface area per unit volume, and three different volumes are tested per fiber type, different amounts of polycarboxylate high-range water reducer (HRWR) and cement paste are added to each base mixture during the mixing process to maintain satisfactory fresh properties for adequate consolidation under external vibration. Paste is prepared with the same water to cementitious materials ratio as the base mix design to avoid affecting the strength of the hardened composite. The paste additions are deemed necessary to coat the extra surface area of the fibers and provide adequate fresh properties. Table 2 summarizes mixture proportions of each mixture after the paste and HRWR additions. Cross sections of the broken beams for each mixture show the amount of fibers present and the dispersion of them throughout the concrete matrix (Figure 1).

Test Methods and Performance Parameters

Test Methods

The VKelly test is employed to evaluate the fresh properties of the fiber mixtures designed for the current study. This test measures two parameters: (1) VKelly slump, which indicates the fresh concrete mixture's initial yield stress and is calculated by doubling the Kelly ball penetration into the fresh concrete under its own weight; and (2) VKelly index, which reports how well the mixture responds to vibration for consolidation. The VKelly index is found by timing the Kelly ball's penetration depth into the fresh concrete under controlled external vibrations at 6-s intervals. This is then used to plot the penetration depth versus time$^{0.5}$, the slope of which is taken as the VKelly index. To calculate the VKelly index of each mixture, three test trials are performed and the recorded values are averaged. The VKelly test was originally developed for slip-form paving applications, for which a VKelly index of 0.6–1.2 in./s$^{0.5}$ (15–30 mm/s$^{0.5}$) was deemed suitable. Further details of the VKelly test can be found in Taylor *et al.* (*26*).

ASTM C1609 Standard (*17*) is used to evaluate the flexural performance of the FRC mixtures. Following this standard, third point bending tests are carried out using a 400-kip (1780-kN) capacity Instron universal testing machine with an external data acquisition system connected to two linear variable differential transformers (LVDTs). The support span length on the bottom of the beams is set to 12 in. (305 mm). Two point loads are applied to the top of the beam spaced 4 in. (100 mm) apart and placed symmetrically 4 in. (100 mm) inside of the bottom supports (Figure 2*a*). The supports and top load applicators are made from oval shaped dowel bars to facilitate free rotation at the supports. Because the supports of the testing apparatus are restrained against translational movements, some lateral friction forces can be generated between the specimen and the bottom supports as the deflection increases during the test. This may cause some pseudo deflection hardening behavior, especially at high deflection values. However, because all specimens are tested under the same condition, consistent comparisons are made for different fiber types and dosages based on the data collected.

Aluminum tabs are glued to each specimen at the mid-span on both sides of the beam. Magnetic stands are used to hold LVDTs under each tab to measure the mid-span deflection (Figure 2*b*). The deflections obtained from both sides of the beam are averaged to calculate the flexural performance parameters. Following ASTM C1609 Standard, the mechanical load is applied in a displacement-controlled manner at a loading rate of 0.003 in. (0.075 mm) per minute up to a deflection of $L/900$, where L is the support span length. After this

Table 2. Adjusted Mix Proportions after Paste and HRWR Additions

Mix ID	Cementitious materials (lb/yd^3 [kg/m^3])			Aggregates (lb/yd^3 [kg/m^3])		Water (lb/yd^3 [kg/m^3])	HRWR (oz/cwt)	Fiber volume (%)
	Type I/II Cement	Class F Fly Ash	Total	Coarse	Fine			
Base mix	593 (352)	254 (151)	847 (503)	1534 (910)	1257 (745)	322 (191)	0.0	0.0
PP0.5	593 (352)	254 (151)	847 (503)	1533 (909)	1256 (744)	322 (191)	0.6	0.5
PP1.0	599 (355)	257 (152)	856 (507)	1511 (896)	1239 (735)	325 (193)	1.8	1.0
PP1.5	607 (360)	260 (154)	867 (514)	1474 (874)	1208 (716)	330 (196)	3.1	1.5
PVA0.5	610 (362)	261 (155)	871 (517)	1505 (892)	1234 (732)	331 (196)	2.9	0.5
PVA1.0	636 (377)	272 (161)	908 (538)	1455 (863)	1193 (707)	346 (205)	2.8	1.0
PVA1.5	664 (394)	284 (168)	948 (562)	1397 (828)	1144 (678)	361 (214)	6.3	1.5
ARG0.5	593 (352)	254 (151)	847 (503)	1533 (909)	1256 (745)	322 (191)	1.0	0.5
ARG1.0	589 (349)	252 (149)	841 (498)	1525 (904)	1250 (741)	320 (190)	2.6	1.0
ARG1.5	594 (352)	255 (151)	849 (503)	1506 (893)	1234 (732)	333 (197)	2.6	1.5

Note: cwt = hundredweight (100 lb).

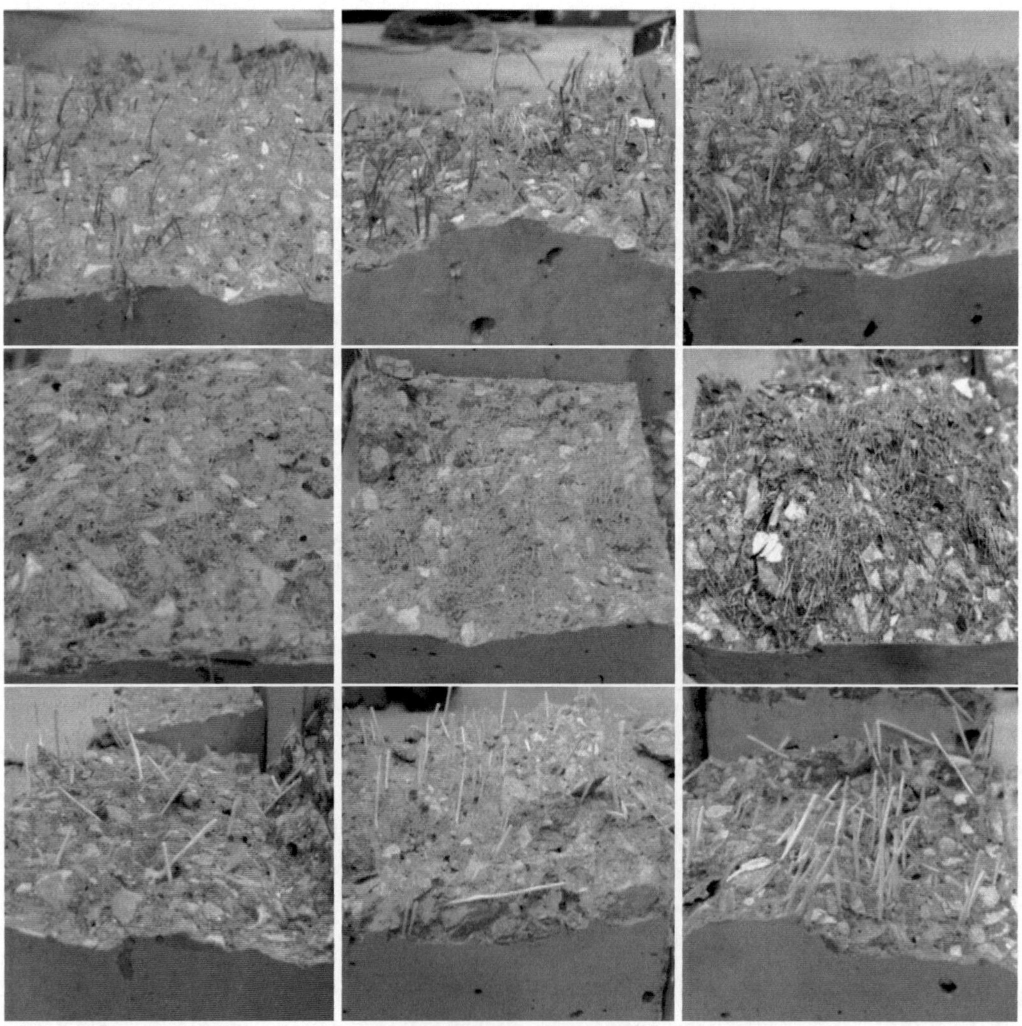

Figure 1. A close view of fiber dispersion in broken beams made with 0.5%, 1.0%, and 1.5% (from left to right) of PP, PVA, and ARG fibers (from top to bottom).

deflection is reached, the loading rate is increased to 0.005 in. (0.127 mm) per minute for the remainder of the test. In the current study, three beams sized 4 × 4 × 14 in. (100 × 100 × 350 mm) are cast for each fiber type and dosage, making a total of 27 specimens. The beam molds are oiled before casting and the fresh concrete is consolidated using a vibrating table. Preferential alignment of the fibers during casting is avoided to the extent possible. The specimens are demolded after 24 h and allowed to cure at 73ºF (23ºC) and 100% relative humidity. All the beams are tested at 7 days of age.

Flexural Performance Parameters

According to ASTM C1609 Standard, a set of measures can be formulated to compare the performance of various specimens based on the load versus deflection curves obtained from the test. The strength in the specimen at any deflection value is calculated using the specimen's dimensions as well as the load resisted at the corresponding deflection by

$$f = \frac{PL}{bd^2} \qquad (1)$$

where

f = strength (psi, MPa),
P = load (lbf, N),
L = span length (in., mm), 12 in. (305 mm) for this study,
b = specimen width (in., mm), 4 in. (100 mm) for this study, and
d = specimen depth (in., mm), 4 in. (100 mm) for this study.

Equation 1 is employed to find the modulus of rupture, which can be considered as a measure of the beam's flexural strength. The modulus of rupture is very nearly the same as the first peak flexural strength, f_1, associated with the load immediately before crack formation, P_1. At deflections larger than that corresponding to the first peak load, the residual load carrying capacity is provided by the fibers in the mixture. Residual load parameters of P_{600} and P_{150} for deflections of $L/600$ and $L/150$, respectively, are used to compute residual strength parameters, f_{600} and f_{150}, using Equation 1, with the load at the corresponding deflections. Residual strength refers to the ability of the FRC specimen to sustain load after cracking has occurred.

In addition, toughness up to a net deflection of $L/150$ (T_{150}) is obtained by calculating the area under the load–deflection curve up to the corresponding deflection (i.e., 0.08 in. [2 mm] for the beam dimensions used in this study). Toughness, which is defined as the energy absorbed by the specimen during the loading process, can be considered as an important parameter for evaluating the flexural performance of FRC mixtures. This is primarily because it quantifies the post-crack load carrying capacity of the specimen through the entire specified range of deflections.

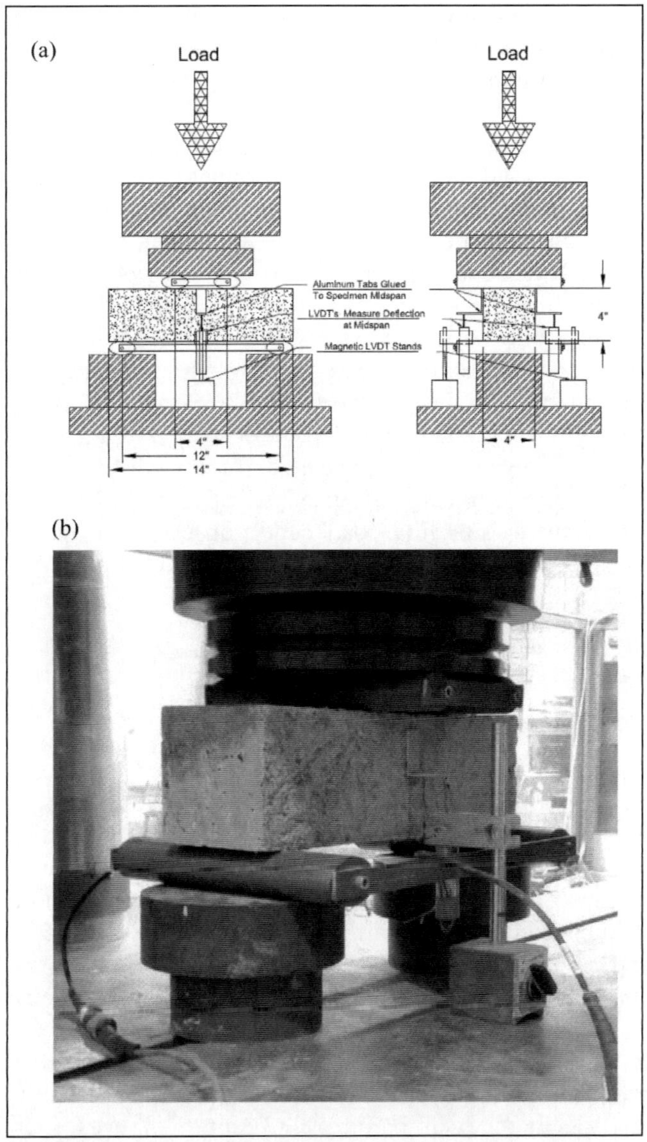

Figure 2. Flexural test setup: (*a*) main dimensions; and (*b*) full configuration.

Finally, the equivalent flexural strength ratio is computed

$$R_{T150} = \frac{150 T_{150}}{f_1 b d^2} \qquad (2)$$

where

R_{T150} = equivalent flexural strength ratio (%),
f_1 = first peak strength (psi, MPa),
T_{150} = Toughness up to $L/150$ deflection (in.-lb, Joules),

b = specimen width (in., mm), 4 in. (100 mm) for this study, and

d = specimen depth (in., mm), 4 in. (100 mm) for this study.

The equivalent flexural strength ratio is a parameter that relates the first peak flexural strength to the toughness of the concrete. A low first peak strength and high toughness would result in a high R_{T150} value, whereas a high first peak strength and low toughness would result in a low R_{T150} value. Because this study utilizes the same concrete matrix for all mixtures, it is expected that R_{T150} values follow the same trend as the toughness values for each mixture. Using the performance parameters from ASTM C1609 Standard, three-specimen average and standard deviation are calculated for the performance parameters of interest. The analysis of such parameters along with the original load–deflection curves provides an in-depth insight into the promise of each fiber type and dosage.

Results and Discussion

Fresh Properties

During the process of mixing and casting the FRC specimens, each fiber type is found to affect the fresh properties of the mixture differently. Table 3 shows the percentage of paste and HRWR dosage added to each mixture along with the corresponding VKelly slump and index. Results show that the PP fibers mix well at 0.5% and 1.0% volumes with a relatively low effect on workability, which can be compensated for by small additions of HRWR. At the 1.5% volume of PP fibers, however, the mixture becomes considerably stiff with an insufficient response to vibration, as reflected in the relatively low VKelly index of 0.13 in./s$^{0.5}$ (3.3 mm/s$^{0.5}$), even after the addition of paste and HRWR.

The PVA fibers show certain limitations in the fresh state. At 0.5% volume, they mix very well with adequate dispersion and workability. However, the response to vibration is so high that the ball moves through the concrete and no reading can be taken past 6 s. This is the reason for the absence of a reported value for the VKelly index of PVA0.5 mixture in Table 3. For the 1.0% and 1.5% PVA mixtures, clumping of the fibers becomes an issue, as once the volume percentage of fibers exceeds a certain limit, the fibers come together and form 1- to 2-in. diameter clumps that contain only sand, fiber, and paste (Figure 3). Because continued mixing would not disperse the clumps, the specimens are cast and consolidated with the existing clumps. Although fiber dispersion admixtures like methylcellulose, latex, and acrylic are reported to help solve this issue (*27*), it was decided to not utilize these admixtures because they were not needed for PP and ARG mixtures, in addition to their lack of practicality in full-scale projects.

ARG fibers demonstrate the smallest effect on the fresh properties of the mixture, which is expected because of their relatively low aspect ratio, as well as their specific gravity being closest to that of the concrete matrix. The ARG fiber mixtures respond well to vibration, as can be inferred from their comparatively high VKelly indices. The VKelly indices of the mixtures tested in this study are all below 0.6–1.2 in./s$^{0.5}$ (15–30 mm/s$^{0.5}$), which is the VKelly index recommended for slip-form paving. This indicates that fiber addition has a strong effect on the fresh properties. Although further research is needed to determine the optimal VKelly parameters for FRC mixtures, the VKelly index values reported in this study can be effectively used for capturing the relations between fiber types, dosages, and corresponding fresh properties.

Flexural Test Results

The beams made with the fiber types and dosages considered in the current study all show a deflection hardening behavior, immediately after the formation of the first

Table 3. Fresh Properties of FRC Mixtures

			VKelly slump		VKelly index	
Mix ID	% Paste Added	HRWR (oz/cwt)	Average (in. [mm])	COV (%)	Average (in./s$^{0.5}$ [mm/s$^{0.5}$])	COV (%)
PP0.5	0.0	0.6	1.5 (38)	6.4	0.47 (11.9)	5.8
PP1.0	2.5	1.8	1.3 (33)	7.1	0.33 (8.4)	3.5
PP1.5	6.7	3.1	1.0 (25)	10.9	0.13 (3.3)	12.6
PVA0.5	4.9	2.9	2.6 (66)	15.4	-	-
PVA1.0	13.0	2.8	1.3 (33)	7.4	0.25 (6.4)	9.6
PVA1.5	23.1	6.3	2.0 (51)	8.2	0.19 (4.8)	12.6
ARG0.5	0.0	1.0	2.3 (58)	4.2	0.60 (15.5)	3.0
ARG1.0	0.0	2.6	1.9 (48)	13.4	0.43 (10.9)	3.1
ARG1.5	2.0	2.6	1.3 (33)	7.4	0.30 (7.6)	5.4

Note: cwt = hundredweight (100 lb); COV = coefficient of variation; NA = not available.

Figure 3. Cross section of broken PVA FRC beam with lumps of re-aggregated fibers.

crack. The first crack is characterized by the first sharp drop in the load–deflection curve. A positive slope following this drop indicates the deflection hardening behavior. For the specimens containing PVA fibers, this initial post-crack deflection hardening response is followed by a deflection softening behavior. One of the PVA0.5 beams did not have a completed data set because the load dropped past 20% of the initial value and the safety stop was triggered on the testing machine. The safety stop was adjusted to a lower value for subsequent tests. One of the ARG1.5 specimens also did not have a completed data set because a crack formed directly under the portion of the beam where the aluminum tab was glued and the tab became disconnected.

All volumes tested for PP and ARG fibers show a continued deflection hardening behavior after the formation of the first crack. It is interesting to note the presence of a second sudden drop in some of the load–deflection curves, which is an indicator of a second crack forming in the specimen. This can be noticed for some of the ARG1.0 and ARG1.5 specimens, as well as one of the PP1.5 specimens. The presence of multiple cracks in the ARG1.0, ARG1.5, and PP1.5 specimens demonstrates the fiber's role in transferring load across the crack, helping develop the full strength of the matrix. All other mixtures listed in Table 3 are found to fail from a single crack that continues to propagate. Figure 4 shows the failed specimens with single and double cracking patterns for comparison. The load versus mid-span deflection curves obtained for the beams with 0.5% and 1.5% fiber dosages are shown in Figure 5.

The average and standard deviation of five performance parameters introduced by ASTM C1609 Standard are extracted from load–deflection curves. Figure 6*a* contains the comparison of the three-specimen average first peak strength, which is a measure of the composite flexural strength before cracking. It is observed that for PP and ARG fibers, the first peak strength increases with the increase of fiber dosage. For PP fibers, this increase is small (close to 2%) when comparing the PP0.5 and PP1.0 specimens. However, the increase is found to approach 8% when the PP1.5 specimens are tested. For ARG fiber specimens, a larger increase (14.5%) in the first peak strength is recorded between the ARG0.5 and ARG1.0 specimens, but the increase becomes less significant (1.5%) when the ARG1.5 specimens are tested.

A 5.2% average decrease in the first peak strength occurrs between the PVA0.5 and PVA1.0 specimens, but a 10.6% average increase occurrs between the PVA1.0 and PVA1.5 specimens. The inconsistencies in the PVA fiber mixtures can be attributed to the lack of uniform dispersion in the PVA1.0 and PVA1.5 mixtures, as the fiber clumps may have created weak spots in the beam. For the 0.5% fiber volume mixtures, PVA specimens show the highest first peak strength followed by PP and finally ARG fibers. This trend is reversed for the 1.0% volume mixtures and continues for the 1.5% volume mixtures. Because of high standard deviations in the PP0.5, ARG0.5, PP1.5 and ARG1.5 data, it is uncertain as to which fiber provides a larger first peak strength. However, it can be generalized that increasing the volume of both PP and ARG macrofibers can slightly increase the first peak strength of the beams in the range of fiber volumes tested.

Figures 6*b* and 6*c* show the three-specimen average and standard deviation for $L/600$ and $L/150$ residual strength of each mixture, respectively. It is clear that for all three fibers tested, increasing the fiber volume increases the residual strength. ARG specimens consistently provide the highest $L/600$ residual strength,

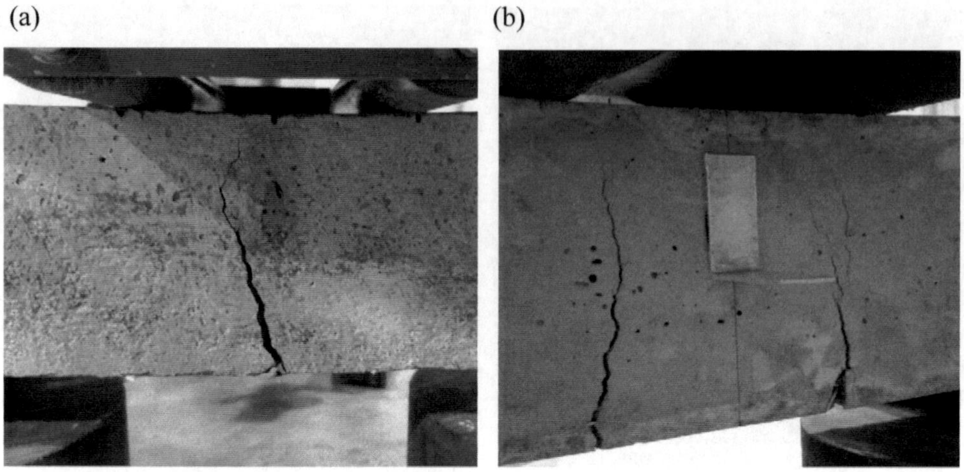

Figure 4. A close view of (*a*) single; and (*b*) double crack pattern failure.

whereas PP specimens show the lowest for 0.5% volume and PVA specimens show the lowest for 1.5% volume. The increase in the *L*/600 residual strength with increasing fiber volume is more pronounced for PP fiber specimens than for PVA fiber specimens. Large increases occur in the *L*/600 residual strength with increasing PP and ARG fiber content, especially between the 0.5% and 1.0% fiber volumes. The PP1.0 specimen experiences a 148.4% increase over PP0.5, whereas the ARG1.0 specimen witnesses a 194.1% increase over ARG0.5. These increases become less pronounced with the increase of dosage to 1.5% for all the fiber types, especially for ARG (4.0%). However, the increase still remains significant for PP1.5 (66.0%). The PVA specimens show a more linear increase in the *L*/600 residual strength across all three fiber volumes tested, but the increase is still over 20% higher between PVA0.5 and PVA1.0 than between PVA1.0 and PVA1.5. The *L*/150 residual strengths shown in Figure 6*c* clearly indicate improvements with increasing fiber volumes for all three fibers tested, but the ARG1.0 mixture is found to be exceptionally high and PVA1.0 is found to be relatively low for the *L*/150 residual strength. PVA fiber mixtures also consistently show the lowest *L*/150 residual strengths, because of deflection softening behavior at large deflections for all fiber volumes, whereas ARG fiber mixtures consistently show the highest *L*/150 residual strengths, because of the fibers' ability to sustain loads at large deflections.

Toughness is calculated as the area under the load-deflection curve up to *L*/150 deflection, as per ASTM C1609 Standard. Figure 7*a* shows the three-specimen average and standard deviations of toughness. For all the fibers tested, toughness increases with increasing the fiber volume percentage. The toughness of PP and PVA fiber mixtures show a linear increase with increasing fiber volume. ARG fiber mixtures, however, show a much larger increase in toughness from ARG0.5 to ARG1.0 (143.1%) compared with ARG1.0 to ARG1.5 (7.6%). PVA and ARG fiber mixtures consistently provide the lowest and highest toughness values, respectively. The three-specimen average and standard deviation of equivalent flexural strength ratio are shown in Figure 7*b*. The equivalent flexural strength ratio follows the same trend as toughness.

The post-crack performance parameters, including f_{600}, f_{150}, T_{150}, and R_{T150} demonstrate a consistent increase with increasing the fiber volume for all the three fibers tested. PP results indicate linear increases with fiber volume for all the four post-crack parameters investigated. ARG1.0 specimens are found to generate exceptional results compared with PVA1.0 and PP1.0 specimens. ARG1.0 results are also well above the linear trend between ARG0.5 and ARG1.5. The results indicate a marginal benefit for flexural performance by increasing the ARG fiber percentage from 1.0% to 1.5% for the mix design used in this study. PVA fiber mixtures generally follow a linear increasing trend for post-crack parameters, despite the lack of consistent dispersion and difficulties encountered while mixing the fibers.

The fact that the fiber with the largest cross section and lowest aspect ratio shows the best flexural results in this study is counterintuitive, given that a higher aspect ratio and smaller diameter result in more surface area per volume of fibers in contact with the concrete matrix. PP fibers have the highest aspect ratio and similar length to the ARG fibers, but did not perform as well as the ARG fibers, meaning that the ability of the ARG fibers to form a bond with the concrete matrix is superior to

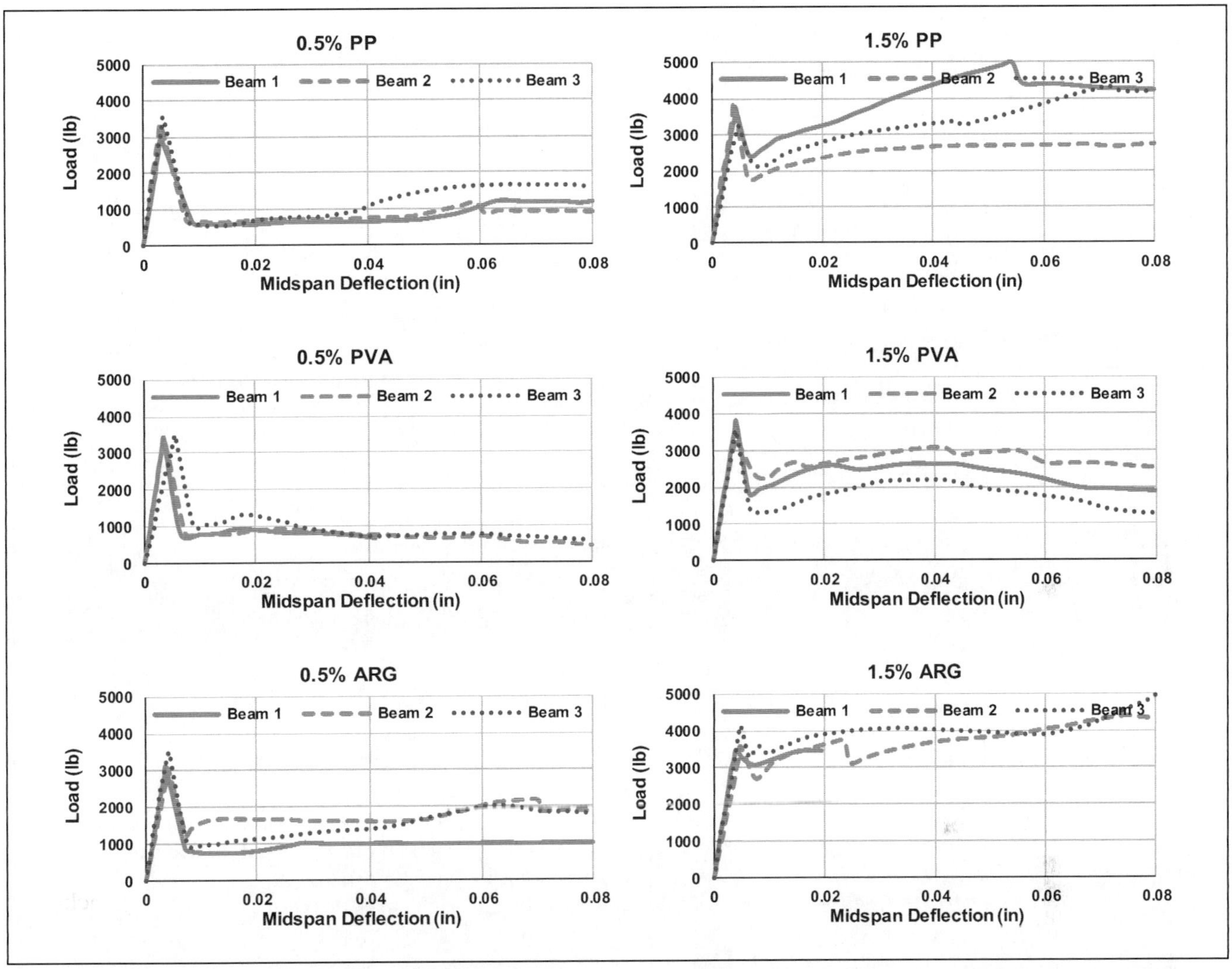

Figure 5. Load–deflection curves obtained from flexural tests on the specimens made with PP, PVA, and ARG fibers (1 in. = 25.4 mm; 1 lb = 4.45 N).

that of the PP fibers. The shorter PVA fibers that rank the second highest aspect ratio have half of the length of the PP and ARG fibers. This means that there are more PVA fibers per unit volume of concrete, but the fibers have less surface area available per fiber to bond with the concrete and develop the expected strength. Initially, the shorter PVA fibers were selected because of the reported ability to form a strong bond with concrete, but the results indicate that the longer PVA fibers may have been more effective in providing flexural residual strength and toughness. For all the three types of fiber, the failure mechanism is characterized by the fibers pulling out from the concrete matrix, rather than rupturing. This indicates that the limiting factor contributing to flexural and tensile performance is the bond strength between the fiber and the concrete matrix. A higher strength concrete matrix may have been able to develop more of the fiber strength and increase toughness. However, for the FRC mixtures tested in this study, the obtained results indicate that long fibers with superior ability to resist pull-out from the concrete matrix are most effective to increase post-crack flexural properties.

Conclusions

This research effort focused on comparing the flexural performance of FRC mixtures made with three synthetic macrofibers with comparable aspect ratios (i.e., PP, PVA, and ARG fibers), dispersed at volume percentages of 0.5%, 1.0%, and 1.5%. The results of this work showed that the VKelly test qualifies as a suitable test for fresh properties of FRC, as it can capture static and dynamic responses that are particularly important for fiber mixtures.

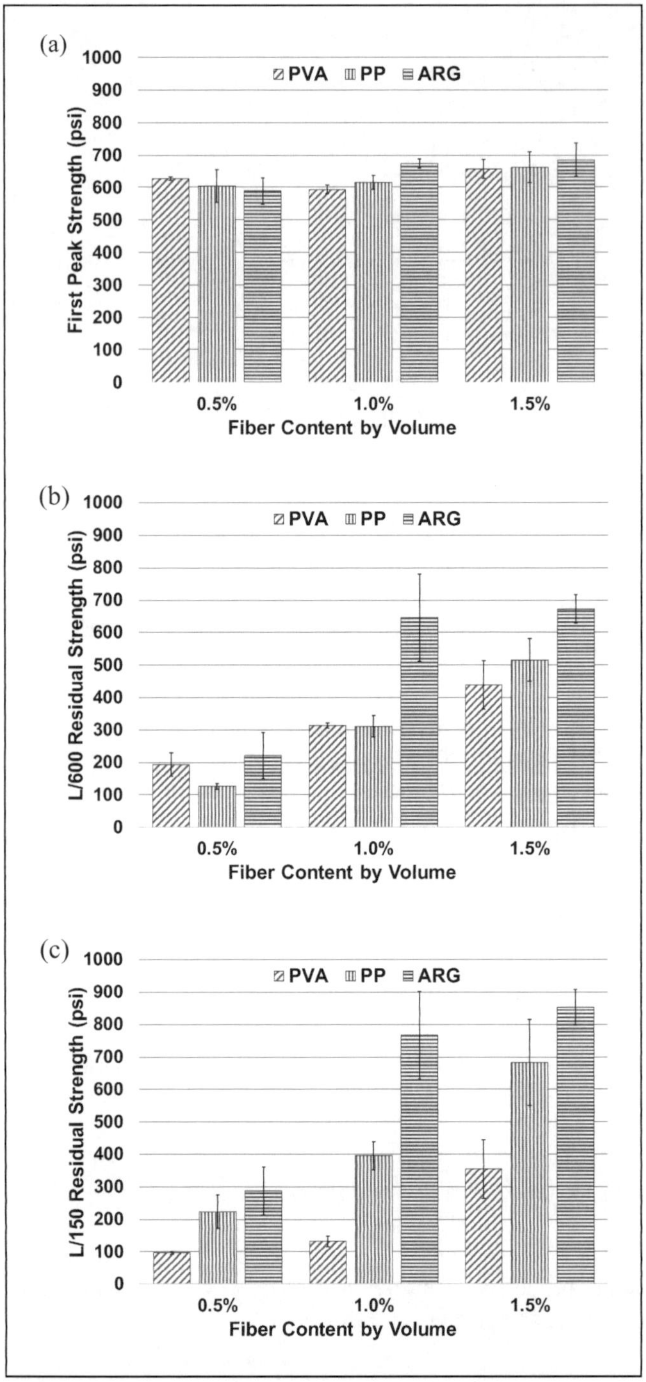

Figure 6. Comparisons of the average and standard deviation of three flexural performance measures: (*a*) first peak strength; (*b*) *L*/600 deflection residual strength; and (c) *L*/150 deflection residual strength (1 psi = 0.0069 MPa).

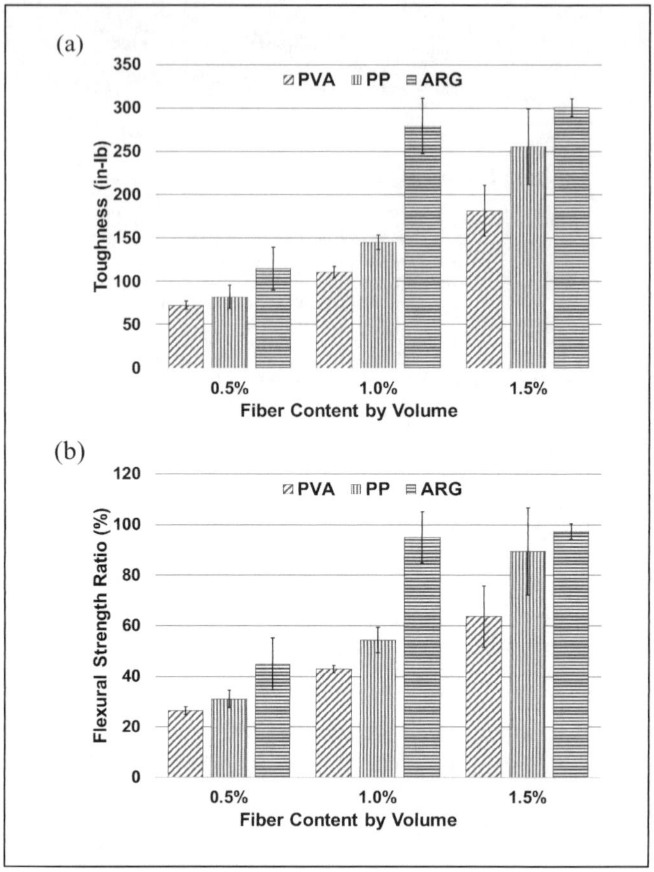

Figure 7. Comparisons of (*a*) toughness and (*b*) equivalent flexural strength ratio obtained for the three fibers tested in this study (1 in.-lb = 0.113 Joules).

The investigation of flexural performance of the tested FRC beams indicate that:

- Increasing fiber volumes for PP and ARG macrofibers slightly increases the first peak strength of the composite, but this trend was less pronounced and cannot be ascertained based on the results of this study for the PVA macrofibers tested, most likely because of weak spots in the cross section as a result of clumps formed during mixing.
- Increasing the fiber volume between 0.5% and 1.5% resulted in increased post-crack performance parameters of residual strength, toughness, and equivalent flexural strength ratio for the three fibers tested.
- ARG fibers consistently produced the highest residual strength and toughness values of the three fibers tested, with a more pronounced advantage for the 1.0% volume ARG fiber mixture. Small increases in all flexural parameters were noticed between the ARG1.0 and ARG1.5 mixtures for the concrete matrix used in this study. ARG fibers had the smallest effect on the fresh properties of the concrete at each volume tested.
- PVA fibers consistently produced the lowest toughness values for the three volume percentages tested. When PVA fiber volumes were increased to 1.0% or more, the fibers tended to re-aggregate

and form clumps with the sand and paste, creating a non-homogenous mixture that had apparent effects on the composite pre-crack flexural performance, but less of an effect on the composite post-crack flexural performance.

- PP fibers showed the most consistent increases in post-crack flexural properties when fiber volumes were increased. Although the toughness and residual strength of PP mixtures were lower than those of ARG mixtures at all volumes, the PP0.5 versus ARG0.5 and PP1.5 versus ARG1.5 mixtures showed comparable flexural performance results. PP fibers mixed well at 0.5% and 1.0% volume, but the PP1.5 mixture had low vibrational response and would require a high HRWR dosage to achieve consolidation during large-scale placements, especially if rebar is present.
- A higher aspect ratio did not necessarily translate to higher flexural performance for the fibers tested in the current study. ARG fibers had the lowest aspect ratio, yet showed the highest flexural performance, mainly because of the ability of the resin coating on the fibers and helical twists to form a strong bond with the concrete.

Acknowledgments

The research study for which the results are reported in this manuscript was sponsored by Iowa Department of Transportation as well as Accelerated Bridge Construction University Transportation Center (ABC-UTC) at Florida International University with Iowa State University and University of Nevada, Reno as partner universities. ABC-UTC is a Tier 1 UTC funded by the U.S. Department of Transportation. Authors would like to acknowledge and thank the sponsors for their support. Authors would like to thank the contributions of Owens Corning, ReforceTech, Forta, and Kuraray for their generous donation of fibers. The assistance of lab personnel, Bob Steffes and Jeremy McIntyre, as well as Doug Wood and Owen Steffens is appreciated as well.

Author Contributions

The authors confirm contribution to the paper as follows: study conception and design: Dr. Behrouz Shafei, Dr. Peter Taylor, and Dr. Brent Phares; laboratory tests and data collection: Mr. Michael Dopko; analysis and interpretation of results: Mr. Michael Dopko, Dr. Meysam Najimi, and Dr. Xuhao Wang; draft manuscript preparation: Mr. Michael Dopko. All authors reviewed the results and approved the final version of the manuscript.

References

1. Banthia, N., and S. M. Soleimani. Flexural Response of Hybrid Fiber-Reinforced Cementitious Composites. *Materials Journal*, Vol. 102, No. 6, 2005, pp. 382–389.
2. Thomas, J., and A. Ramaswamy. Mechanical Properties of Steel Fiber-Reinforced Concrete. *Journal of Materials in Civil Engineering*, Vol. 19, No. 5, 2007, pp. 385–392.
3. Bentur, A., and S. Mindess. *Fibre Reinforced Cementitious Composites*. CRC Press, Boca Raton, FL, 2006.
4. Zheng, Z., and D. Feldman. Synthetic Fibre-Reinforced Concrete. *Progress in Polymer Science*, Vol. 20, No. 2, 1995, pp. 185–210.
5. ACI Committee 544. *Report on Fiber Reinforced Concrete, ACI 544.1R-96 (Reapproved 2009)*. American Concrete Institute, Farmington Hills, MI, 2009.
6. Alani, A. M., and D. Beckett. Mechanical Properties of a Large Scale Synthetic Fibre Reinforced Concrete Ground Slab. *Construction and Building Materials*, Vol. 41, 2013, pp. 335–344.
7. Patnaik, A., S. Bajaj, J. Lewis, J. Payer, and R. Liang. Retardation of Strength Degradation of Reinforced Concrete Due to Steel Bar Corrosion with Fiber Additions. *Proc., DOD Virtual Corrosion Conference*, University of Akron, Akron, OH, 2013, pp. 16–17.
8. Dopko, M., M. Najimi, B. Shafei, X. Wang, P. Taylor, and B. Phares. Flexural Performance Evaluation of Fiber Reinforced Concrete Incorporating Multiple Macro-Synthetic Fibers. Presented at 97th Annual Meeting of the Transportation Research Board, Washington, D.C., 2018.
9. Dopko, M., M. Najimi, B. Shafei, X. Wang, P. Taylor, and B. Phares. Assessment of Strength and Restrained Shrinkage Behaviors on Micro-Carbon Fiber Reinforced Concrete with Binary Chemical Admixtures. Presented at 97th Annual Meeting of the Transportation Research Board, Washington, D.C., 2018.
10. Mohod, M. V. Performance of Polypropylene Fibre Reinforced Concrete. *Journal of Mechanical and Civil Engineering*, Vol. 12, No. 1, 2015, pp. 28–36.
11. Patnaik, A. *MiniBar Reinforced Concrete (MRC)*. Report RFT-AP-MB-R03. University of Akron, Akron, OH, 2013.
12. Hamoush, S., T. Abu-Lebdeh, and T. Cummins. Deflection Behavior of Concrete Beams Reinforced with PVA Micro-Fibers. *Construction and Building Materials*, Vol. 24, No. 11, 2010, pp. 2285–2293.
13. Buratti, N., C. Mazzotti, and M. Savoia. Post-Cracking Behavior of Steel and Macro-Synthetic Fibre-Reinforced Concretes. *Construction and Building Materials*, Vol. 25, No. 5, 2011, pp. 2713–2722.
14. Shafiq, N., T. Ayub, and S. U. Khan. Investigating the Performance of PVA and Basalt Fibre Reinforced Beams Subjected to Flexural Action. *Composite Structures*, Vol. 153, 2016, pp. 30–41.
15. Roque, R., N. Kim, B. Kim, and G. Lopp. *Durability of Fiber-Reinforced Concrete in Florida Environments*. University of Florida, Tallahassee, FL, 2009.
16. Fraternali, F., V. Ciancia, R. Chechile, G. Rizzano, L. Feo, and L. Incarnato. Experimental Study of the Thermo-Mechanical Properties of Recycled PET Fiber-Reinforced Concrete. *Composite Structures*, Vol. 93, No. 9, 2011, pp. 2368–2374.
17. ASTM International. *Standard Test Method for Flexural Performance of Fiber-Reinforced Concrete (Using Beam*

with Third-Point Loading). ASTM, C 1609/C 1609M-05. ASTM International, West Conshohocken, PA, 2005.
18. Banthia, N., and R. Gupta. Influence of Polypropylene Fiber Geometry on Plastic Shrinkage Cracking in Concrete. *Cement and Concrete Research*, Vol. 36, No. 7, 2006, pp. 1263–1267.
19. Studinka, J. B. Asbestos Substitution in the Fibre Cement Industry. *International Journal of Cement Composites and Lightweight Concrete*, Vol. 11, No. 2, 1989, pp. 73–78.
20. Betterman, L. R., C. Ouyang, and S. P. Shah. Fiber-Matrix Interaction in Microfiber-Reinforced Mortar. *Advanced Cement Based Materials*, Vol. 2, No. 2, 1995, pp. 53–61.
21. Li, V. C. *Engineered Cementitious Composites (ECC) Material, Structural, and Durability Performance*. University of Michigan, Ann Arbor, MI, 2008.
22. Mohammadi Mohaghegh, A. *Use of Macro Basalt Fibre Concrete for Marine Applications*. Doctoral dissertation. KTH Royal Institute of Technology, Stockholm, Sweden, 2016.
23. Wu, G., X. Wang, Z. Wu, Z. Dong, and G. Zhang. Durability of Basalt Fibers and Composites in Corrosive Environments. *Journal of Composite Materials*, Vol. 49, No. 7, 2015, pp. 873–887.
24. Patnaik, A., L. Miller, and P. C. Standal. Fiber Reinforced Concrete Made from Basalt FRP Minibar. *Proc., Concrete Innovation Conference*, Oslo, Norway, 2014, pp. 11–13.
25. Johnston, C. D. *Fiber-Reinforced Cements and Concretes*. CRC Press, Boca Raton, FL, 2001.
26. Taylor, P., X. Wang, and X. Wang. *Concrete Pavement Mixture Design and Analysis (MDA): Development and Evaluation of Vibrating Kelly Ball Test (VKelly Test) for the Workability of Concrete*. InTrans Project Report. Iowa State University, Ames, IA, 2015.
27. Cao, J., and D. D. L. Chung. Carbon Fiber Reinforced Cement Mortar Improved by Using Acrylic Dispersion as an Admixture. *Cement and Concrete Research*, Vol. 31, No. 11, 2001, pp. 1633–1637.

The Standing Committee on Basic Research and Emerging Technologies Related to Concrete (AFN10) peer-reviewed this paper (18-06370).

Opinions and conclusions expressed in this manuscript are of the authors and do not necessarily represent those of sponsors.

Article

TRR
JOURNAL OF THE TRANSPORTATION RESEARCH BOARD

Transportation Research Record
2018, Vol. 2672(27) 13–23
© National Academy of Sciences:
Transportation Research Board 2018
Article reuse guidelines:
sagepub.com/journals-permissions
DOI: 10.1177/0361198118773571
journals.sagepub.com/home/trr
SAGE

Examining Curing Efficiency using Neutron Radiography

Mehdi Khanzadeh Moradllo[1], Luca Montanari[1], Prannoy Suraneni[1], Steven R. Reese[2], and Jason Weiss[1]

Abstract

Many state highway agencies use prescriptive specifications for the curing of concrete bridge decks, pavements, flatwork, or structural elements. For example, concrete pavements are frequently specified to have a curing compound applied shortly after placement and bridge decks typically require seven days of wet curing. These specifications are often based on historical practices that have developed over the last century as opposed to quantitative measurements of performance. New approaches to curing are being introduced which include advanced curing compound formulations or internal curing, for example. However, clear information is not always available as to how this may affect curing requirements. This paper demonstrates the potential to use neutron radiography to quantify the degree of hydration at various distances from the finished surface. It describes how different curing approaches affect cement hydration in terms of both time and distance from the surface. The results show that in a sample exposed to drying after one day the top 12.5 mm (1/2 in) of the mortar was dramatically affected by evaporation, and the degree of hydration in this region was 32% lower than in a 14-day moist-cured sample. Also, the use of superabsorbent polymers increased degree of hydration by about 3.7–7.8% for sealed samples and samples exposed to drying. While these results are preliminary, it is believed by the authors that neutron radiography provides a powerful approach that could be used to determine equivalent curing requirements for new materials.

State highway agencies (SHAs) specify a series of steps to be taken after concrete is placed to ensure that the concrete is cured. Curing consists of maintaining "satisfactory moisture content and temperature of concrete for a sufficient period of time during and immediately following placing so that desired properties may develop" (*1*). Curing is an important step in developing long-lasting concrete since it promotes hydration of the cement leading to strength and durability. Curing in practice largely consists of supplying additional water through water ponding or wet fabrics (e.g., burlap) or providing a curing compound that minimizes water loss from the system through a plastic membrane or a membrane-forming curing compound (*2*). Because of its importance in concrete technology, much has been written about curing (*3, 4*).

Many specifications provide a required duration during which curing should be provided. For example, AASTHO recommends three days of curing for pavements and seven days for bridge decks, which extends to ten days when supplementary cementitious materials are used (*5*). Poole found that the duration of curing required was frequently based on attaining some fraction of overall strength (*6*). For example, ACI 301 recommends a minimum curing duration for the concrete to attain 70% of the specified compressive strength which often corresponds to 7-day curing (*7*). Australian specifications relate curing requirements with the environment to which the concrete will be exposed (*8*). Carino and Clifton identified that the required curing depends on the type of concrete being used, and performed research to examine the curing needed for high-performance concrete (*9*). Other researchers have examined the quality of the materials used to cure concrete. Vandenbossche reviewed curing compounds and application techniques used by Minnesota Department of Transportation (*10*). Hajibabaee et al. examined the influence of different curing compound compositions on the curling of paste samples (*11*).

[1]School of Civil and Construction Engineering, Oregon State University, Corvallis, OR
[2]School of Nuclear Science and Engineering, Oregon State University, Corvallis, OR

Corresponding Author:
Address correspondence to Jason Weiss: jason.weiss@oregonstate.edu

Recently, the concept of internal curing (IC) has been explored. IC is provided by partially replacing the fine aggregates with an equivalent volume of pre-wetted fine lightweight aggregates (*12, 13*). Others have internally cured concrete using superabsorbent polymers (SAP) that absorb water (pore fluid) quickly during mixing and release it slowly into the hydrating cement paste which then maintains high relative humidities over time (*14*). Relatively little work has been performed to understand how the duration of curing may change when internally cured mixtures are used. However, work has examined the use of IC for pervious mixtures (*15*) and IC mixtures have shown great promise in field applications (*16–18*). This work will examine the potential impact of using internally cured concrete with differing external curing environments.

It is important to note that while the entire cross-section of concrete being placed requires curing, the surface of the concrete is the most sensitive to moisture loss, and therefore is the most sensitive to curing (*3*). Cather defined a zone at the surface of concrete (i.e., the skin of the concrete), called the curing-affected zone (CAZ) (*19*). While estimates of the size of the CAZ may vary, the CAZ can be thought of as extending from the surface to a depth of approximately a half inch to an inch and a half (12.5–37.5 mm).

A variety of test methods have been used to assess curing efficiency by either measuring moisture/moisture loss or surface/near-surface physical properties. The simplest test appears to be to visually verify that the surface of the concrete remains wet (*20*). Menzel provided early work to examine the influence of evaporation rates on concrete (*21*). Jensen and Hansen proposed the use of a novel curing sensor that consisted of a capillary tube that is placed on the concrete surface (*22*). Hanson measured the relative humidity (RH) near the surface of concrete during drying (*23*); however, it should be noted that currently many of the RH sensors are relatively large (*24*) and have a larger variability at high RH levels. Powers et al. studied how the duration of wet curing pastes of differing water-to-cement ratio (w/c) affected the permeability of the paste (*25*). Surface absorption was used measure the effectiveness of water curing (*26*). Others used a variety of methods to measure permeability, water absorption, RH, abrasion resistance, hardness of the surface concrete, and also a notched cylindrical test specimen to measure the tensile strength of concrete at different depths (*27, 28*). Poole asked whether it may be a "better approach" to "rely totally on performance testing," citing that curing could be ceased once concrete achieved a specific strength level (*6*).

In the current study, a non-destructive technique (in which the sample can be tested repeatedly over time at different conditions) is used to measure the quantity of water that has reacted with the cementitious material. This non-destructive technique is called neutron radiography (NR) (*29, 30*). NR consists of passing a neutron beam through a material, in this case a cementitious mortar. A portion of this neutron beam passes through the sample while a portion is absorbed or reflected. The natural logarithm of the ratio of the intensity of the neutron that is transmitted through the concrete (I_T) and the original beam (I_O) is related to an absorption coefficient for the sample (μ_s) and the thickness of the sample (x_s) as in (*31*)

$$\ln\left(\frac{I_T}{I_O}\right) = -\mu_S x_S \qquad (1)$$

Unlike X-ray radiography, which is based on the density of a material, NR is primarily related to the neutron atomic cross-section (a term used to express the likelihood that an incident neutron will interact with the target nucleus). The neutron atomic scatter and absorption cross-sections are unique to each atom, but are particularly large for the first isotope of hydrogen (H-1), which comprises 99.9885% of all hydrogen found in nature (*32*). For example, H-1 has a total scattering cross-section of 82.02 barns as compared with other materials frequently found in concrete that are an order of magnitude lower (e.g., sodium 3.28, calcium 2.90, magnesium 4.03, oxygen 4.23, and silicon 2.17) (*32*). As a result, NR is used to quantify water content in materials and has been widely used for fuel cell testing (*33*).

NR has been used to study cementitious materials. Wedding et al. provided early use of NR in studies of microcracking in concrete (*34*). De Beer and co-workers studied the water movement during the drying process of concrete (*35*). Villani et al. investigated drying in concrete containing shrinkage reducing admixtures (*36*). Wyrzykowski et al. used neutron tomography to examine the loss of water from fresh concrete (*37*). NR has been used to investigate fluid absorption in cementitious materials, especially at sawcuts and cracks (*38–41*). Zhang et al. and Kanematsu et al. examined fluid transport and the role of cracking on fluid transport (*42–44*). Jones et al. employed neutron tomography to quantify the crack filling ability of polymers in cracked concrete (*45*). Trtik et al. used neutron tomography to examine the movement of water from lightweight aggregate (*46*) and SAP (*47*) for IC applications. Livingston has focused on using neutron tomography to measure deleterious reactions like delayed ettringite formation (*48*). While the vast majority of the aforementioned studies have focused on the qualitative movement of "free water" in concrete, Lucero et al. quantified the attenuation coefficients of both the constituent and hydrated materials (*40*). This paper describes the potential use of NR to quantify the water that has reacted with hydration products and to utilize the spatial resolution capabilities of the technique to quantify curing effectiveness.

Research Objective

Although curing is related to wind speed, RH, concrete temperature, and air temperature, this paper focuses only on the impact of a single environment on curing (23 ± 1°C, 50 ± 2% RH). This paper has three objectives. First, this paper examines whether it is feasible to use NR to quantify the degree of hydration (DOH). Second, this paper examines the role of curing on the DOH, particularly in the CAZ. Third, this paper examines the role of SAP in increasing the DOH.

Experimental Program

This paper investigated a series of experiments to evaluate whether NR could be used to measure the DOH and the impact of different practices on curing efficiency. The DOH was measured using isothermal calorimetry on mortars, non-evaporable water based on loss on ignition (LOI) (*49*), and NR. The latter two measurements were performed on sealed samples that were oven dried at 105°C at ages from 6 h to 14 days to remove the evaporable water. The evaluation of curing efficiency was measured on samples that were water ponded in a curing room or had wet burlap applied for mortar mixtures with and without SAP. The samples in each of these conditions were exposed to drying at prescribed times as shown in Table 1 and were oven dried at 14 days to remove the evaporable water. The following section describes the details of the sample program, conditioning, and testing.

Materials and Mixture Proportions

Two mortar mixtures were prepared, a plain mortar mixture and a mortar mixture containing SAP using the proportions given in Table 2.

A Type I ordinary Portland cement conforming to ASTM C 150 was used in this study. The chemical composition of the cement from the mill sheet was CaO 63.20%, SiO_2 20.40%, Al_2O_3 4.70%, Fe_2O_3 3.30%, SO_3 2.80%, MgO 2.50%, Na_2O_{eq} 0.54% and a LOI of 1.30%. The fine aggregate is a natural river sand with a specific gravity of 2.49, and an absorption of 3.10%. A commercially available SAP used in the experiments.

The amount of SAP for IC was chosen to provide a volume of water that was equivalent to the volume of chemical shrinkage developed by the cement, as described elsewhere (*50, 51*). The fluid absorption capacity of the

Table 1. Curing Regimes for Prismatic and Cubic Samples

		Prismatic samples (7.6 × 7.6 × 30.0 cm)			Cubic samples (5.1 × 5.1 × 5.1 cm)
Mixture	Curing age (days)	Water ponded	Burlap then drying	Sealed	Sealed
OPC	0.25				√***
	0.5		√**		√
	1		√		√
	3	√*	√		√
	7	√	√		√
	14	√		√***	√
SAP	0.25				√
	0.5				√
	1		√		√
	3	√	√		√
	7	√	√		√
	14	√		√	√

Note: *Duration of wet curing; **The age that burlap was removed and sample was exposed to 50% RH until age of 14 days; ***Duration of sealed curing.

Table 2. Mortar Mixture Proportions (in SSD conditions)

Mixture	w/c	Cement (kg/m^3, m^3/m^3)	Water (kg/m^3, m^3/m^3)	Aggregate (kg/m^3, m^3/m^3)	IC Water (kg/m^3, m^3/m^3)	Dry SAP (kg/m^3)
OPC	0.42	676.6, 0.215	284.2, 0.285	1242.4, 0.500	0	0
SAP	0.42	644.2, 0.205	270.6, 0.271	1188.5, 0.480	45.1, 0.045	1.675

SAP was determined using a pore solution extracted from a cement paste with the same w/c of the mortar mixture 30 minutes after mixing. The fresh paste was placed in a pore solution extractor with nitrogen gas providing a pressure of 200 kPa for a period of 5 minutes as described in Montanari (*51*). The extractor used a cellulose membrane filter with an average pore size of 0.45 μm to minimize the potential for contamination of the solution from cement particles. The extracted pore solution was sealed in a glass vial and stored at 5°C to minimize the potential for carbonation. The absorption capacity of the SAP was measured using the "teabag method" described in Montanari (*51*) and an absorption of 26.9 (g/g) was determined.

Casting and Curing

The mixtures were prepared following a procedure similar to ASTM C 305. However, after mixing the cement with water and adding the dry aggregates, the SAP was introduced into the mixture in a dry state. An additional 120 s of stirring were added to the standard mixing steps, after the introduction of the dry SAP, to ensure a good dispersion of the polymers. Water in the mixture accounted for the absorption from aggregate.

Three sample geometries were prepared. Cubic samples, $5.1 \times 5.1 \times 5.1\ cm^3$, were prepared for determination of DOH through LOI and NR. Prismatic samples ($7.6 \times 7.6 \times 30.0\ cm^3$) were cast to investigate the impact of different practices on curing efficiency by NR.

These samples were prepared by tamping and vibrating the fresh mortar. The sealed samples were placed under plastic during curing and then in two thick plastic bags 6 h after casting. These samples were maintained in a sealed condition at 23 ± 1°C until their time of testing (6 h to 14 days) using LOI or NR.

The prismatic samples were demolded after 12 h from casting and then sealed with aluminum tape on all sides but the finished surface. Three different curing regimes were implemented for curing prism samples (described in Table 1): (1) water ponded in moist room at 23 ± 1°C for 3 days, 7 days, and 14 days; (2) wet curing with saturated wet burlap for different periods of time and then drying at 50 ± 2% RH and 23 ± 1°C until age of 14 days; and (3) sealed curing for 14 days at 23 ± 1°C. Wet-cured samples in moist room were stored in a sealed condition at 23 ± 1°C after termination of the curing until age of 14 days in order to test all samples at the same age.

Experimental Methods

Determining the Non-Evaporable Water Content using LOI

After achieving the required age, mortar samples were crushed into smaller pieces using a hammer and were placed in an oven at 105°C for 3 days. After drying in the oven, approximately 10 g of crushed pieces were placed in a furnace at 1000°C for a minimum of 6 h. The cement powder and the fine aggregate were dried in an oven at 105°C and then ignited in a furnace at 1000°C, in order to correct non-evaporable water content based on the LOI of the cement and aggregate. The non-evaporable water content was calculated as the difference between the 1000 and 105°C mass measurements for the crushed pieces. Then, this measurement was corrected for the LOI of the cement powder and the fine aggregate (*49*). The DOH of the sample was found by dividing the measured non-evaporable water content of the sample with the amount of non-evaporable water in the same sample when it is fully hydrated, which is assumed 0.23 g/g cement (*49, 52*).

Isothermal Calorimetry

The mortars were tested using isothermal calorimetry. After mixing, approximately 10 g of the mortar was poured into a glass ampoule, and then tamped to ensure good consolidation of a sample. The glass ampoule was then sealed with a lid to prevent the evaporation of any moisture during the experiment. Next, the samples were placed in an isothermal calorimeter (TA Instruments), which was preconditioned at 23°C. The cumulative heat release was measured for the mortar samples until 7 days after mixing. For all samples, testing was performed in duplicate, and the results are presented as the average of the values from the two samples. The difference in the cumulative heat release between the internally cured sample and the non-internally cured sample is used to calculate the change in the DOH in these samples.

Neutron Radiography (NR)

After the curing period, 1-inch (≍25 mm) slices were cut from cubic and prismatic samples using a wet saw for the NR experiment. The samples were then dried at 105°C to remove the evaporable water. Neutron imaging was conducted at the Neutron Imaging Facility (NRF) at Oregon State University (OSU) Radiation Center. This facility utilizes a 1.1 MW water-cooled research reactor that uses uranium/zirconium hydride fuel elements in a circular grid array. The beam within the NRF has a collimation ratio (L/D) of 115 ± 4 and a thermal neutron flux of $9.4 \times 10^5 \pm 1.6 \times 10^4\ cm^{-2}\ s^{-1}$. The NRF at OSU uses an image intensifier coupled with a light-tight right angle adapter to a charge-coupled device (CCD) camera as a neutron detection scheme. The scintillation screen is composed of 5 μm diameter needle cesium iodide crystals doped with Gd. A Nikon D610 camera with 50 mm f/1.2 lens was used to capture images

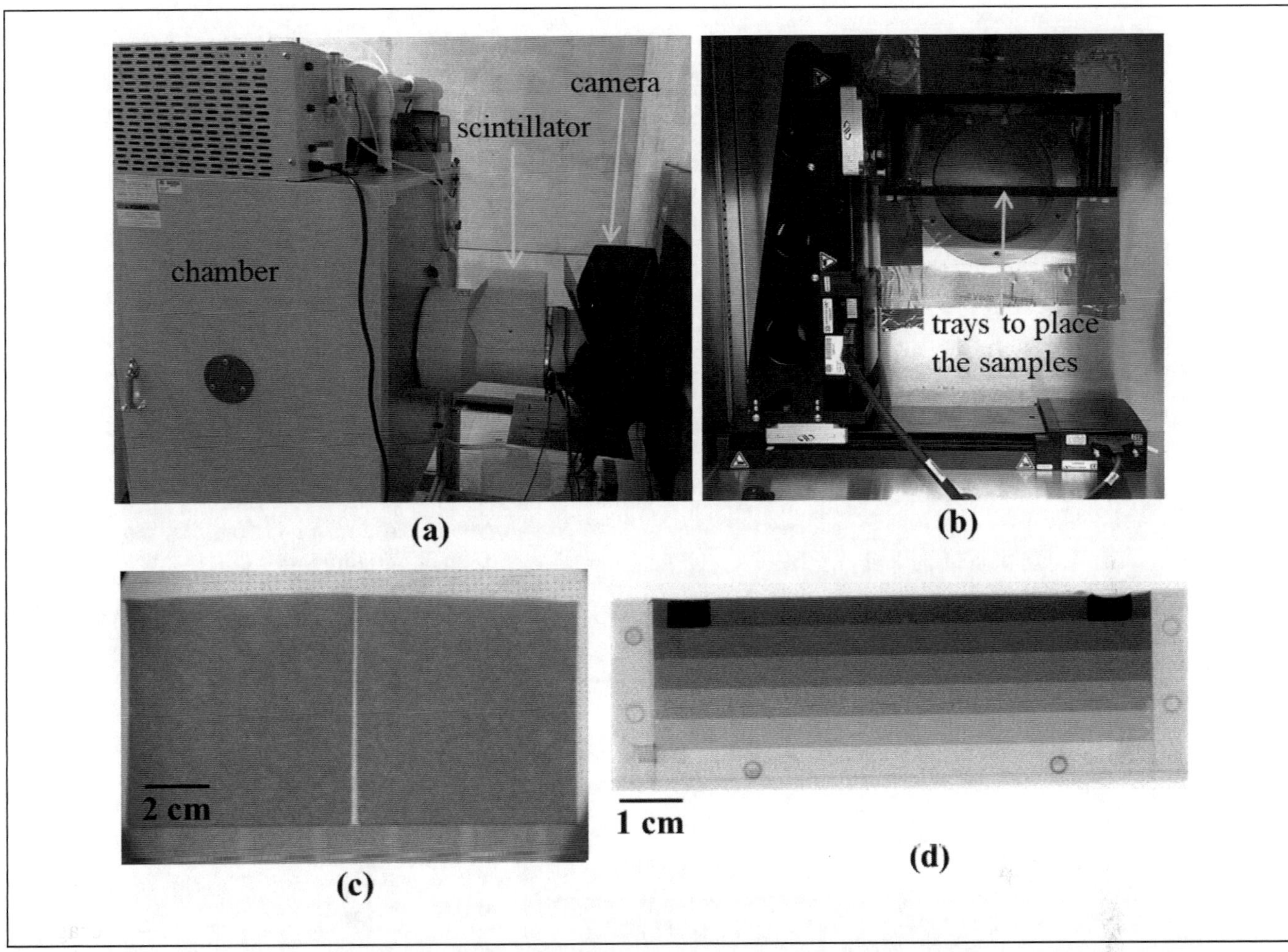

Figure 1. Experimental set up and typical image from NR; (*a*) Chamber with scintillator at the back. Neutron beam is parallel to the scintillator, (*b*) Translating stage inside the chamber, (*c*) Raw images from camera, (*d*) Water-filled stepped cells to measure water attenuation coefficient, thickness 1–5 mm.

through the scintillator with a spatial resolution of approximately 90 μm. The field of view (FOV) is approximately 150 mm $\times$ 150 mm. Three images of each sample were taken with an exposure time of 2 s. A motorized translating stage was used to image multiple samples (Figure 1). A median filter was used to combine three images of a sample to eliminate the artifacts caused by the gamma rays. In addition, multiple images were taken from the background (detector's FOV) with open beam (flat-field image) and with closed beam (dark-field image) for the background correction. The images from NR were processed and analyzed with ImageJ software. This process includes a background correction and calculation of intensity in the region of interest. The attenuation coefficient of water (μ_W = 0.1267 mm^{-1}) was calculated by scanning the empty and water-filled stepped cells (Figure 1*d*) and developing a relationship between water thickness (x_w) and optical density (OD = $\ln(I/I_0)$) based on Equation 1 (*31*). In addition, the attenuation coefficients of the oven-dried fine aggregate (μ_A = 0.0179 mm^{-1}) and unhydrated cement powder (μ_C = 0.0159 mm^{-1}) were determined using Equation 1. The thicknesses of these materials were corrected by their estimated packing densities. The following section describes the details of DOH calculation from NR measurements.

Equation 1 can be re-written to account for the different phases present in a hydrated cement paste:

$$\ln\left(\frac{I_T}{I_O}\right) = -\mu_S x_S = \sum_{i=1}^{n} -(\mu_i V_i)x_S$$
$$= -(\mu_A V_A + \mu_C V_C + \mu_{GS} V_{GS} + \mu_{GW} V_{GW} + \mu_{CW} V_{CW} + \mu_{CS} V_{CS})x_S \quad (2)$$

where μ_i is the attenuation coefficient and V_i is the volume fraction of component "*i*" where the components are referred to as aggregate (A), unhydrated cement (C),

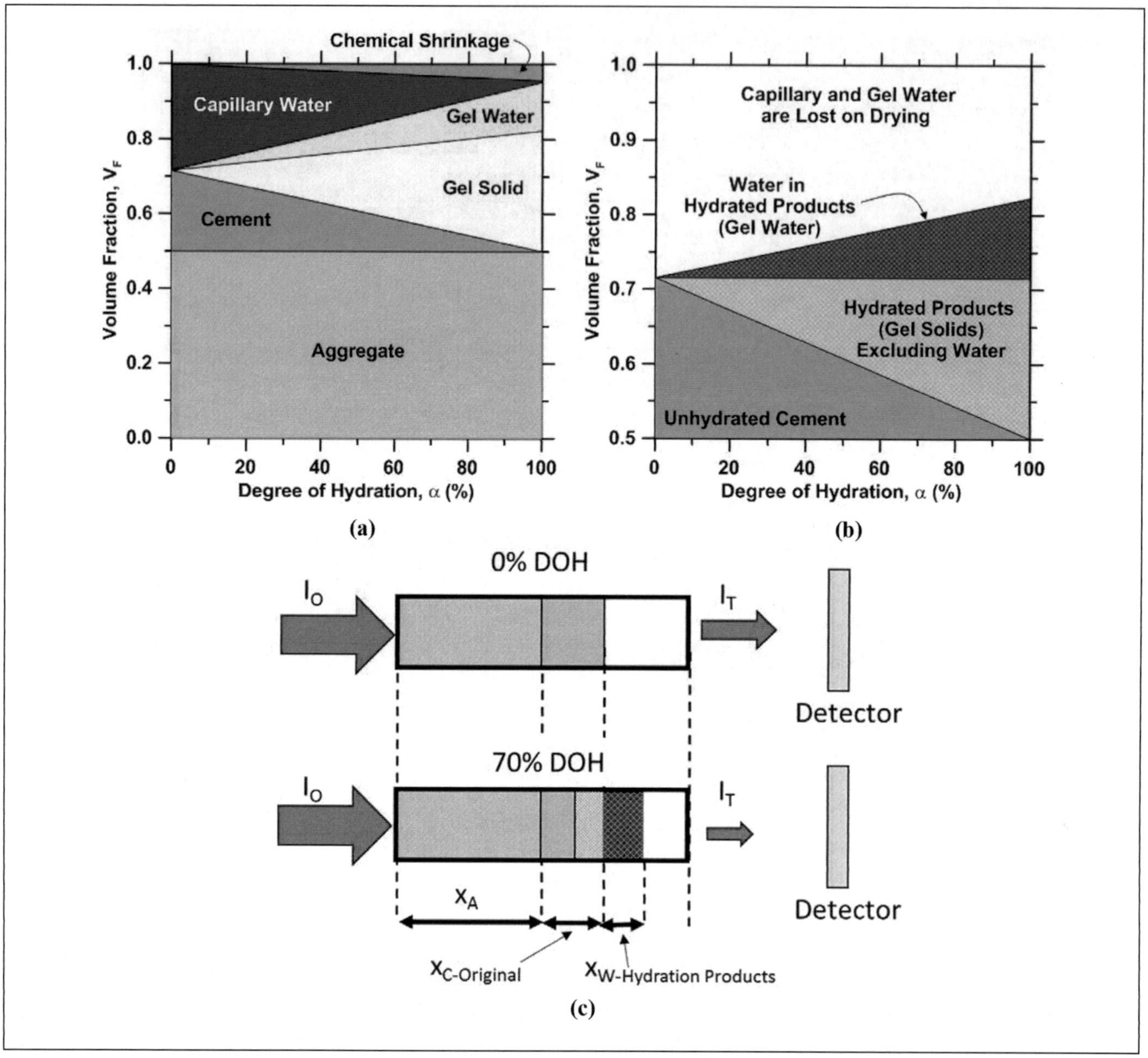

Figure 2. (*a*) Volumetric proportions of the raw materials and reacted phases following the Powers and Brownyard model, (*b*) a volumetric examination of the water bound in the hydrated products, and (*c*) a schematic representation of materials volume for use in Equations 1 and 3.

hydrated gel solids (GS), gel water (GW), capillary water (CW) and chemical shrinkage (CS). In this formulation, the phase volumes vary as a function of the DOH. It should be noted that this formulation does not include entrained or entrapped air; however, this could be added. Figure 2 shows an example of the volumetric proportions of the raw materials and reacted phases plotted against DOH following the Powers and Brownyard model (*48*) for a mortar with w/c of 0.42.

The change in beam intensity after interacting with a sample can be examined in a different manner, as shown in Figure 2*b*. One can remove the water components that are lost on heating to 105°C (CS, CW, GW) and consider that the volume of the aggregate and cement do not change during the test (i.e., cement and aggregate are not lost from the system). In that case one can assume that a portion of gel solid is the original cement volume ($V_{\text{c-Original}}$) (hydrated products excluding water and the water in the hydrated product) and a portion is the water that has reacted with the cement (non-evaporable water, $V_{\text{W-Hydration Products}}$). In doing so, the attenuation coefficient can be determined as

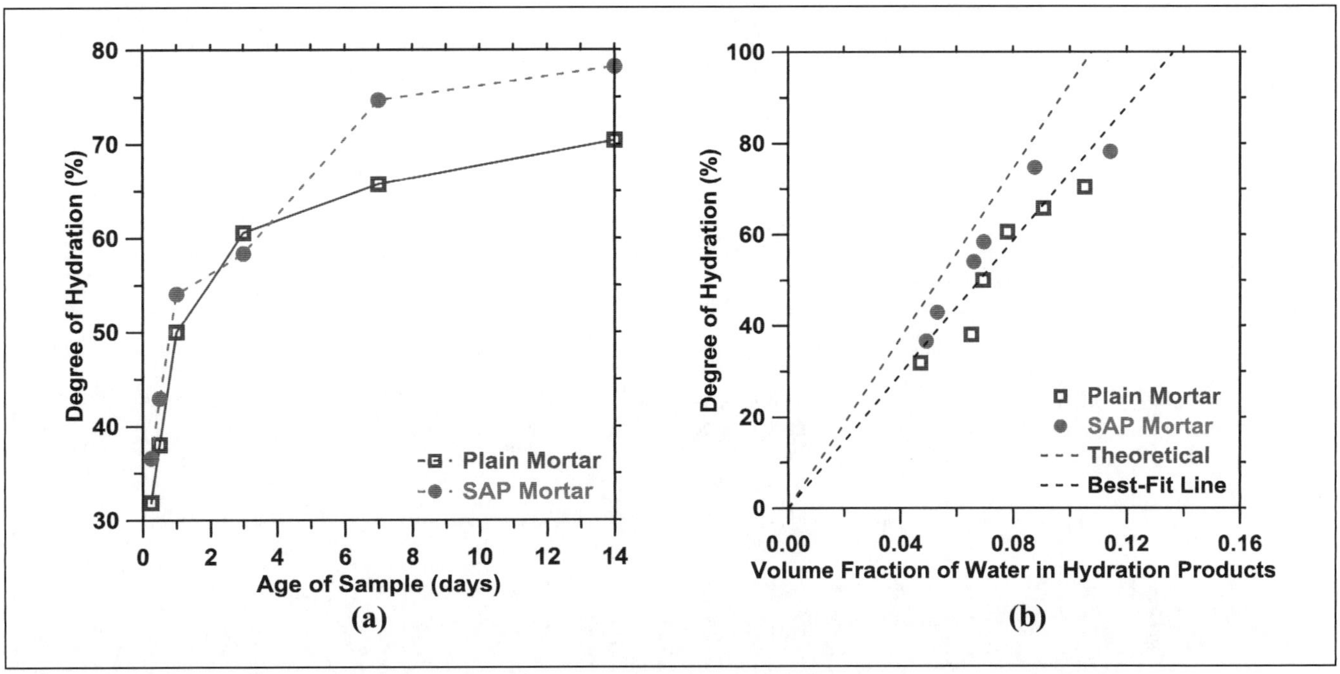

Figure 3. (*a*) Calculated DOH over time by LOI method and (*b*) DOH vs. volume fraction of water in hydrated materials by NR.

$$\ln\left(\frac{I_T}{I_O}\right) = -\left(\mu_A V_A + \mu_C V_{C-\mathrm{Original}} + \mu_W V_{W-\mathrm{Hydration\ Products}}\right) x_S \quad (3)$$

Equation 3 can be rearranged to solve for the volume of the water in the hydrated products by NR:

$$V_{W-\mathrm{Hydration\ Products}} = -\left(\frac{1}{\mu_W}\right)\left(\frac{\ln\left(\frac{I_T}{I_O}\right)}{x_S} - \mu_A V_A - \mu_C V_{C-\mathrm{Original}}\right) \quad (4)$$

Experimental Results and Discussion

The non-evaporable water content (as measured on sealed samples by LOI) can be used to determine the DOH, as plotted in Figure 3*a*. The DOH increases over time as may be expected. The DOH of the mortars with SAP is similar to mortars without SAP at early ages; however, the mixture with SAP has a higher DOH at 14 days (7.8%). This is comparable to the increase in the DOH measured by isothermal calorimetry at 7 days, which is 7.2%. The DOH from LOI is plotted against the volume fraction of water in the hydrated materials (Equation 4) obtained using NR for sealed samples in Figure 3*b*. A linear relationship exists between the DOH and the volume fraction of water in the hydrated products as expected (*53*) (Figure 2). This water in the hydrated products was determined by calculating the water that chemically reacted with the cement (i.e., the original water content minus the capillary and gel water). Therefore, the following equation was used to determine the DOH from the calculated fraction of water in hydrated products that is determined using Equation 4:

$$DOH = C_1 V_{W-\mathrm{Hydration\ Products}} \quad (5)$$

where C_1 is a constant predicted using the Powers and Brownyard model (930 %/unitless). The best-fit line of the experimental data corresponds to approximately 15% more water in the hydrated products (783 %/unitless) than that predicted using Powers and Brownyard model (Figure 3*b*). Although the exact reason for this discrepancy is not known, it may be as a result of differences in modern cements from those used by Powers and Brownyard, including changes in cement chemistry, the inclusion of limestone, and grinding aids. Further, this is predicated based on the attenuations of each of the constituent materials and the accuracy of the non-evaporable water content that decreases with tests on mortar. Nevertheless, Equation 5 can be a powerful tool to convert neutron radiograph measurements to the DOH.

Figure 4 shows the gradient in the DOH as a function of depth for plain and SAP samples exposed to differing external curing environments. The DOH values were calculated using Equations 4 and 5. The calculated C_1 (783 %/unitless) was used to develop Figure 4. Based on Figure 4, *a* and *b*, wet curing with saturated burlap showed a significant improvement in DOH of both plain and SAP systems at the top 20 mm of the sample surface.

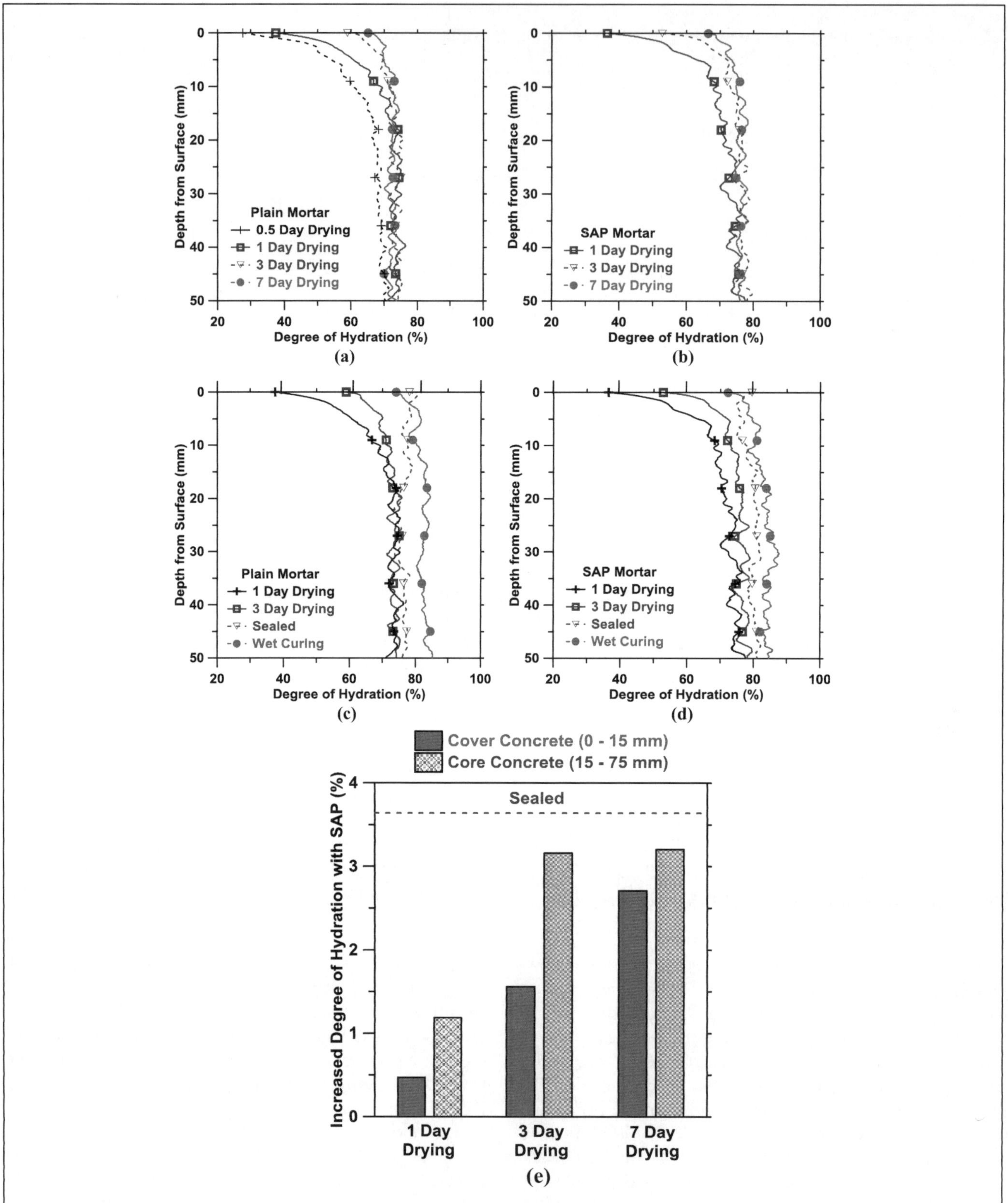

Figure 4. DOH over depth by NR for (*a*) plain mortar exposed to drying, (*b*) SAP mortar exposed to drying, (*c*) comparison between various curing practices—plain mixture, (*d*) comparison between various curing practices—SAP mixture (symbols are at every 100th data point), and (e) increase in DOH over depth by implementing SAP. (Note: "wet curing" represents water-ponded curing in moist room and "drying" shows wet curing with saturated wet burlap for different periods of time and then drying at 50 ± 2% RH and 23 ± 1°C until age of 14 days).

For example, the DOH increased about 24% in 5 mm of the surface of the plain and SAP samples by extending the wet burlap duration from 1 to 3 days. The sealed plain samples had 3.2% greater hydration at the core than the samples exposed to drying at 1 day, whereas the water-cured samples had 12.5% greater hydration at the core than the samples exposed to drying at 1 day (Figure 4, *c* and *d*). In addition, the sealed and 14-day water-cured SAP samples showed 8.5% and 14% higher DOH at the core compared with the sample exposed to drying at 1 day.

An increase in DOH in samples containing SAP is presented in Figure 4*e*. Information is segmented into the cover material (CAZ) in the top 15 mm, the core material 15–75 mm, and in sealed mortar. The SAP samples exhibited small changes in DOH when exposed to drying at 1 day compared with the plain system. However, the SAP increased DOH more for samples exposed to drying at 3 and 7 days as well as the sealed samples. In addition, while defining the CAZ is difficult, this work found that the depth of the plain sample at which 70% DOH was obtained at 14 days was 12 mm for exposure to dry at 1 day and 7.5 and 5.5 mm for exposure to drying at 3 and 7 days, respectively. The mixture containing SAP decreased the CAZ by nearly 4 mm (1/6 in) for the 3-day and 7-day exposures to drying.

Future Studies

Although this paper has outlined that NR is an emerging technology that can be used to evaluate curing efficiency, further work is needed to develop the database of information needed to improve curing specifications. For example, the current paper only considered one environmental exposure (23 ± 1°C, 50 ± 2% RH) and evaporation rate. Additional evaporation rates should be considered, especially when the temperature is lower because of the reduced rate of hydration. Further, the current study only considered mixtures made with ordinary Portland cement, and as such the role of supplementary materials should be considered in future work. With respect to the use of NR, the current approach relied on relatively thin samples (25 mm thick); however, thicker samples could be evaluated for mortar (up to 60 mm) and for concrete (up to 75 mm or 100 mm) depending on mixture proportions. It is also recommended that to ensure the sample surface is parallel to the beam, additional steps need to be taken using a rotational table as described by Henkensiefken et al. (*54*).

Conclusion

In conclusion, this paper has shown that NR is an experimental technique that can be used to measure the water contained in hydrated cements and can be used to determine the degree of hydration in a cementitious system. The results have indicated that the volume of water in the hydrated products (water in the gel solids) can be related to the non-evaporable water determined by LOI. This can also be correlated with the degree of hydration as assessed using isothermal calorimetry. NR can quantify the extent of hydration at various distances from the drying surface. It was shown in this work that obtaining the spatial information provides the ability to assess the influence of curing methods on the hydration of cement in concrete. This is particularly useful in determining the CAZ. In the mixture exposed to drying after 1 day, the top 12.5 mm (1/2 inch) of the mortar was dramatically affected by the loss of water to evaporation. Although the top 5 mm of the surface of the sample exposed to drying at 1 day had a degree of hydration that was 32% less than the 14-day moist-cured sample, it should be noted that the overall degree of hydration in the core mortar was 11% less than a water-cured sample (3% less than a sealed sample). This clearly indicates transport of water to and from the core of the concrete (*24, 36*). The use of SAP demonstrated benefits in terms of both an increase in the overall degree of hydration (this can conservatively be assumed to be approximately 3.7–7.8%) and a reduction in the depth of the CAZ by 4 mm (1/6 in) when SAP is used. Further work would be necessary to develop the quantifiable correlation between the use of SAP and equivalent moist-curing durations.

Acknowledgments

The authors gratefully acknowledge support for this work through a gift provided by BASF to the last author in support of research as well as support through the pooled fund study (Project No. U1117C) managed by the Oklahoma Department of Transportation.

Author Contributions

The authors confirm contribution to the paper as follows. Study conception and design: Mehdi Khanzadeh Moradllo, Jason Weiss; data collection: Mehdi Khanzadeh Moradllo, Luca Montanari, Prannoy Suraneni; analysis and interpretation of results: Mehdi Khanzadeh Moradllo, Luca Montanari, Prannoy Suraneni, Jason Weiss; draft manuscript preparation: Mehdi Khanzadeh Moradllo, Luca Montanari, Prannoy Suraneni, Steven Reese, Jason Weiss. All authors reviewed the results and approved the final version of the manuscript.

References

1. Kosmatka, S. H., W. C. Panarese, and B. Kerkhoff. *Design and Control of Concrete Mixtures*. Portland Cement Association, Skokie, Ill., 2011.
2. ASTM C309-11. *Standard Specification for Liquid Membrane-Forming Compounds for Curing Concrete*. American

Society of Testing and Materials, West Conshohocken, Pa., 2015.
3. ACI 308R-16. *Guide to External Curing of Concrete*. American Concrete Institute, Farmington Hills, Mich., 2016.
4. Taylor, P. C. *Curing Concrete*. CRC Press, Boca Raton, Fla., 2013.
5. AASHTO T23. *Standard Method of Test for Making and Curing Concrete Test Specimens in the Field*. American Association of State and Highway Transportation Officials, Washington, D.C., 2017.
6. Poole, T. S. *Guide for Curing of Portland Cement Concrete Pavements*. FHWA Publication FHWA-RD-02-099. Turner-Fairbank Highway Research Center, McLean, Va., 2006.
7. ACI 301R-16. *Specifications for Structural Concrete*. American Concrete Institute, Farmington Hills, Mich., 2016.
8. AS 3600-2009. *Concrete Structures*. Standards Australia, Sydney, Australia, 2009.
9. Carion, N., and J. Clifton. *Outline of a National Plan on High Performance Concrete: Report on the NIST/ACI Workshop*. National Institute of Standards and Technology, Gaithersburg, Md., 2000.
10. Vandenbossche, J. *A Review of the Curing Compounds and Application Techniques Used by the Minnesota Department of Transportation for Concrete Pavements*. Minnesota Department of Transportation, Saint Paul, 1999.
11. Hajibabaee, A., M. Khanzadeh Moradllo, and T. Ley. Comparison of Curing Compounds to Reduce Volume Change from Differential Drying in Concrete Pavement. *International Journal of Pavement Engineering*, 2016, pp. 1–10. http://dx.doi.org/10.1080/10298436.2016.1210442.
12. Bentz, D., and J. Weiss. *Internal Curing: A 2010 State-of-the-Art Review*. NISTIR 7765. National Institute of Standards and Technology, Washington, D.C., 2011.
13. Jensen, O., and K. Kovler. *Internal Curing of Concrete - State-of-the-Art Report of RILEM Technical Committee 196-ICC*. Report rep041. RILEM Publications, 2007.
14. Schroefl, C., V. Mechtcherine, P. Vontobel, J. Hovind, and E. Lehmann. Sorption Kinetics of Superabsorbent Polymers (SAPs) in Fresh Portland Cement-based Pastes Visualized and Quantified by Neutron Radiography and Correlated to the Progress of Cement Hydration. *Cement and Concrete Research*, Vol. 75, 2015, pp. 1–13.
15. Kevern, J., and D. Brown. Internal Curing of Pervious Concrete using Lightweight Aggregate. Pervious Paradise, San Diego, 2014. https://www.slideshare.net/DaleFisher4/internal-curing-of-pervious-concrete-using-lightweight-aggregate. Accessed July 29, 2017.
16. Weiss, J., and D. Morian. Review of Internally Cured Concrete as It Relates to Pavements. Presented at 96th Annual Meeting of the Transportation Research Board, Washington, D.C., 2017.
17. Barrett, T., A. Miller, and J. Weiss. *Documentation of the INDOT Experience and Construction of the Bridge Decks Containing Internal Curing in 2013*. No. FHWA/IN/JTRP-2015/10. Purdue University, West Lafayette, Ind., 2015.
18. Rupnow, T. *Lightweight Aggregate and Internally Cured Concrete (ICC) Prove Their Value on Louisiana Bridge*. Trinity LightWeight, Lafayette, La., 2015. http://www.trinitylightweight.com/index.php/case-studies/internal-curing/item/west-congress. Accessed July 29, 2017.
19. Cather, R. How to Get Better Curing. *The Journal of Concrete Society*, Vol. 26, No. 5, 1992, pp. 22–25.
20. Carrier, R., and P. Cady. Evaluating Effectiveness of Concrete Curing Compounds. *Journal of Materials, American Society for Testing and Materials*, Vol. 5, No. 2, 1970, pp. 294–302.
21. Menzel, M. Causes and Prevention of Crack Development in Plastic Concrete. *Proc., PCA Annual Meeting*, Portland Cement Association, Skokie, Illinois, 1954, pp. 130–136.
22. Jensen, O., and P. Hansen. Autogenous Relative Humidity Change in Silica Fume-Modified Cement Paste. *Advanced Cement Research*, Vol. 7, No. 25, 1995, pp. 33–38.
23. Hanson, J. Effects of Curing and Drying Enviroments on Splitting Tensile Strength of Concrete. *Journal Proceedings*, Vol. 65, No. 7, 1968, pp. 535–543.
24. Villani, C. *Transport Processes in Partially Saturated Concrete: Testing and Liquid Properties*. Purdue University, West Lafayette, Ind., 2014.
25. Powers, T. C., L. E. Copeland, and H. M. Mann. Capillary Continuity or Discontinuity in Cement Pastes. *Portland Cement Association Research Development Laboratories*, Vol. 1, No. 2, 1959, pp. 38–48.
26. Chan, R., P. Dhir, and Y. Hewlett. Near-Surface Characteristics of Concrete: Abrasion Resistance. *Materials and Structures*, Vol. 24, No. 122, 1991, p. 122.
27. Hilsdorf, H., and J. Kropp. *Performance Criteria for Concrete Durability*, 1st ed. CRC Press, London, 1995.
28. Carino, N., and K. Meeks. *Curing of High Performance Concrete: Report of the State-of-the-Art*. National Institute of Standards and Technology, Building and Fire Research Laboratory, Gaithersburg, Md., 2000.
29. Berger, H. Neutron Radiography. *Annual Review of Nuclear Science*, Vol. 21, No. 1, 1971, pp. 335–364.
30. Vontobel, P., E. Lehmann, R. Hassanein, and G. Frei. Neutron Tomography: Method and Applications. *Physica B: Condensed Matter*, Vol. 385–386, 2006, pp. 475–480.
31. Hussey, D., D. Spernjak, A. Weber, R. Mukundan, J. Fairweather, E. Brosha, J. Davey, J. Spendelow, D. Jacobson, and R. Borup. Accurate Measurement of the Through-Plane Water Content of Proton-Exchange Membranes Using Neutron Radiography. *Journal of Applied Physics*, Vol. 112, No. 10, 2012, p. 104906.
32. Sears, V. Neutron Scattering Lengths and Cross Sections. *Neutron News*, Vol. 3, No. 3, 1992, pp. 26–37.
33. Hickner, M., N. Siegel, K. Chen, D. Hussey, D. Jacobson, and M. Arif. In Situ High-Resolution Neutron Radiography of Cross-Sectional Liquid Water Profiles in Proton Exchange Membrane Fuel Cells. *Journal Electrochemical Society*, Vol. 155, No. 4, 2008, pp. B427–B434.
34. Wedding, P., W. Najjar, H. Aderhold, and K. Hover. The Application of Neutron Radiography to the Study of Microcracking in Concrete. *Cement Concrete Aggregates*, Vol. 8, No. 2, 1986, pp. 103–109.
35. De Beer, F., W. Strydom, and E. Griesel. The Drying Process of Concrete: A Neutron Radiography Study. *Applied Radiation and Isotopes*, Vol. 61, No. 4, 2004, pp. 617–623.

36. Villani, C., C. Lucero, D. Bentz, D. Hussey, D. Jacobson, and J. Weiss. Neutron Radiography Evaluation of Drying in Mortars with and without Shrinkage Reducing Admixtures. *ACI Spececial Publication*, 2015, pp. 1–24. James, J. B. Novel Characterization Techniques and Advanced Cementitious Materials. *Proc., American Concrete Institute Fall Meeting 2014*, American Concrete Institute, Washington, D.C., 2014.
37. Wyrzykowski, M., P. Trtik, B. Münch, J. Weiss, P. Vontobel, and P. Lura. Plastic Shrinkage of Mortars with Shrinkage Reducing Admixture and Lightweight Aggregates Studied by Neutron Tomography. *Cement and Concrete Research*, Vol. 73, 2015, pp. 238–245.
38. Li, W., M. Pour-Ghaz, J. Castro, and J. Weiss. Water Absorption and Critical Degree of Saturation Relating to Freeze-Thaw Damage in Concrete Pavement Joints. *Journal of Materials in Civil Engineering*, Vol. 24, 2012, pp. 299–307.
39. Li, W., M. Pour-Ghaz, P. Trtik, M. Wyrzykowski, B. Münch, P. Lura, P. Vontobel, E. Lehmann, and J. Weiss. Using Neutron Radiography to Assess Water Absorption in Air Entrained Mortar. *Construction and Building Materials*, Vol. 110, 2016, pp. 98–105.
40. Lucero, C., D. Bentz, D. Hussey, D. Jacobson, and J. Weiss. Using Neutron Radiography to Quantify Water Transport and the Degree of Saturation in Entrained Air Cement Based Mortar. *Physics Procedia*, Vol. 69, 2015, pp. 542–550.
41. Lucero, C., R. Spragg, D. Bentz, D. Hussey, D. Jacobson, and J. Weiss. Neutron Radiography Measurement of Salt Solution Absorption in Mortar. *ACI Materials Journal*, Vol. 114, 2017, pp. 149–159.
42. Zhang, P., F. Wittmann, T. Zhao, E. Lehmann, and P. Vontobel. Neutron Radiography, a Powerful Method to Determine Time-Dependent Moisture Distributions in Concrete. *Nuclear Engineering and Design*, Vol. 241, No. 12, 2011, pp. 4758–4766.
43. Zhang, P., F. Wittmann, T. Zhao, E. Lehmann, L. Tian, and P. Vontobel. Observation and Quantification of Water Penetration into Strain Hardening Cement-based Composites (SHCC) with Multiple Cracks by Means of Neutron Radiography. *Nuclear Instruments and Methods in Physics Research Section A: Accelerators, Spectrometers, Detectors and Associated Equipment*, Vol. 620, No. 2–3, 2010, pp. 414–420.
44. Kanematsu, M., I. Maruyama, T. Noguchi, H. Iikura, and N. Tsuchiya. Quantification of Water Penetration into Concrete Through Cracks by Neutron Radiography. *Nuclear Instruments and Methods in Physics Research Section A: Accelerators, Spectrometers, Detectors and Associated Equipment*, Vol. 605, No. 1–2, 2009, pp. 154–158.
45. Jones, S., D. Bentz, J. Davis, D. Hussey, D. Jacobson, J. Molloy, and J. Sieber. Measurement and Modeling of the Ability of Crack Fillers to Prevent Chloride Ingress into Mortar. *Cement and Concrete Composities*, Vol. 81, 2017, pp. 109–121.
46. Trtik, P., B. Münch, W. J. Weiss, A. Kaestner, I. Jerjen, L. Josic, E. Lehmann, and P. Lura. Release of Internal Curing Water from Lightweight Aggregates in Cement Paste Investigated by Neutron and X-ray Tomography. *Nuclear Instruments and Methods in Physics Research Section A: Accelerators, Spectrometers, Detectors and Associated Equipment*, Vol. 651, No. 1, 2011, pp. 244–249.
47. Trtik, P., B. Muench, J. Weiss, G. Herth, A. Kaestner, E. Lehmann, and P. Lura. Neutron Tomography Measurements of Water Release From Superabsorbent Polymers in Cement Paste. *International Conference Material Science, Aachen-MATSCI*, Vol. III, 2010, pp. 1–10.
48. Livingston, R. A. Neutron Tomography Measurement of Delayed Ettringite Formation in Concrete. *Proc., 14th ISNDCM Non-Destructive Testing (NDT) net*, Marina del Rey, Calif., 2015. http://slideplayer.com/slide/6190161. Accessed January 1, 2017.
49. Fagerlund, G. *Chemically Bound Water as Measure of Degree of Hydration: Method and Potential Errors*. TVBM; Vol. 3150. Division of Building Materials, Lund University, 2009.
50. Bentz, D. P., and K. A. Snyder. Protected Paste Volume in Concrete: Extension to Internal Curing Using Saturated Lightweight Fine Aggregate. *Cement and Concrete Research*, Vol. 29, No. 11, 1999, pp. 1863–1867.
51. Montanari, L. *Toward a Mixture Design Methodology for Internally Cured Concrete that Considers Filling Specific Pores Sizes Caused by Self-Desiccation*. Oregon State University, 2017.
52. Powers, T. C., and T. L. Brownyard. *Studies of the Physical Properties of Hardened Portland Cement Paste*. Portland Cement Association Research Development Laboratories. Chicago, Ill., 1948.
53. Powers, T. C., and T. L. Brownyard. Studies of the Physical Properties of Hardened Portland Cement Paste Part 3. Theoretical Interpretation of adsorption Data. *Journal of American Concrete Institute*, Vol. 18, No. 469, 1946, pp. 469–504.
54. Henkensiefken, R., T. Nantung, and J. Weiss. Saturated Lightweight Aggregate for Internal Curing in Low w/c Mixtures: Monitoring Water Movement Using X-Ray Absorption. *Strain*, Vol. 47, No. Suppl. 1, 2011, pp. e432–e441.

The Standing Committee on Basic Research and Emerging Technologies Related to Concrete (AFN10) peer reviewed this paper (18-01811), which was selected by the Concrete Materials Section (AFN00) to receive its 2018 Best Paper Award.

Article

TRR
JOURNAL OF THE TRANSPORTATION RESEARCH BOARD

Transportation Research Record
2018, Vol. 2672(27) 24–36

DOI: 10.1177/0361198118774157
journals.sagepub.com/home/trr
SAGE

Finite-Element Modeling and Analysis of Early-Age Cracking Risk of Cast-In-Place Concrete Culverts

Yalin Liu[1], Anton K. Schindler[1], and James S. Davidson[1]

Abstract

Extensive cracking was found in several cast-in-place concrete culverts in Alabama. This condition can decrease the long-term durability of the culverts. Early-age stress development in concrete is influenced by temperature changes, modulus of elasticity, stress relaxation, shrinkage, thermal coefficient of expansion, and the degree of restraint. The objective of this study is to determine means to mitigate early-age cracking in culverts by evaluating the cracking risk. Finite-element analysis was used to model the early-age stress by accounting for the following factors: construction sequencing, support restraint, concrete constituents, temperature effects, and the time-dependent development of mechanical properties, creep/relaxation, and drying shrinkage. Experimental results from restraint to volume change tests with rigid cracking frames were used to verify the accuracy of the finite-element analysis. A parametric study was performed to quantify the effect of changing joint spacing, joint type, construction sequence, concrete coefficient of thermal expansion, placement season, and concrete type on the risk of early-age cracking. The finite-element model results revealed that the use of the following measures will reduce the risk of early-age cracking in cast-in-place concrete culverts: concrete with lower coefficient of thermal expansion, contraction joints, sand-lightweight concrete or all-lightweight concrete, and scheduling the casting of the culvert wall to minimize the difference in its placement time relative to its previously cast base. Alternatively, to minimize the contribution of thermal effects on risk of cracking, the construction schedule should be developed to avoid concrete placement during hot weather conditions.

Extensive cracking was found in various cast-in-place (CIP) reinforced concrete box culverts located in the Anniston East Bypass (AEB) in Anniston, Alabama (*1*). An example of a 0.08 in. (2 mm) wide crack in a culvert wall is shown in Figure 1*a*. Many cracks in the AEB culverts had widths greater than 0.04 in. (1 mm) and far exceeded 0.012 in. (0.3 mm), which is the tolerable limit for the exposure condition of these culverts in accordance with ACI 224 (*2*). Because of the numerous wide cracks encountered, condition surveys were conducted on other CIP reinforced concrete box culverts throughout Alabama. Fourteen culverts were surveyed and similar cracks were found in most of them, with most of the excessively wide cracks being transverse and located either in the base or in the walls of the culvert (*1*). The lengths of these culverts varied from 122 ft (37.2 m) to 1,005 ft (306.3 m) (*1*). For the 14 surveyed culverts, 11 used construction joints and 3 used contraction joints to connect adjacent culvert sections. These construction joints (tied joints) had deformed reinforcement continuing longitudinally through the joint, which restrains longitudinal contraction and expansion of adjacent culvert sections. The contraction joints used in three projects, in contrast, allow adjacent concrete sections to contract and expand independently. The spacing of these tied or contraction joints between culvert sections ranged from 30 ft (9.1 m) to 53 ft (16 m) (*1*).

The primary cause of cracking was attributed to restraint of thermal and drying shrinkage effects (*1*). Early-age stress development in concrete is influenced by temperature changes, modulus of elasticity, stress relaxation, shrinkage, thermal coefficient of expansion, and the degree of restraint. Cracks occur when tensile stress in the concrete exceeds its tensile strength.

Objectives and Research Methodology

The primary objective of this paper is to determine means to mitigate early-age cracking in CIP culverts. In order to

[1]Department of Civil Engineering, Auburn University, Auburn, AL

Corresponding Author:
Address correspondence to Yalin Liu: yzl0103@auburn.edu

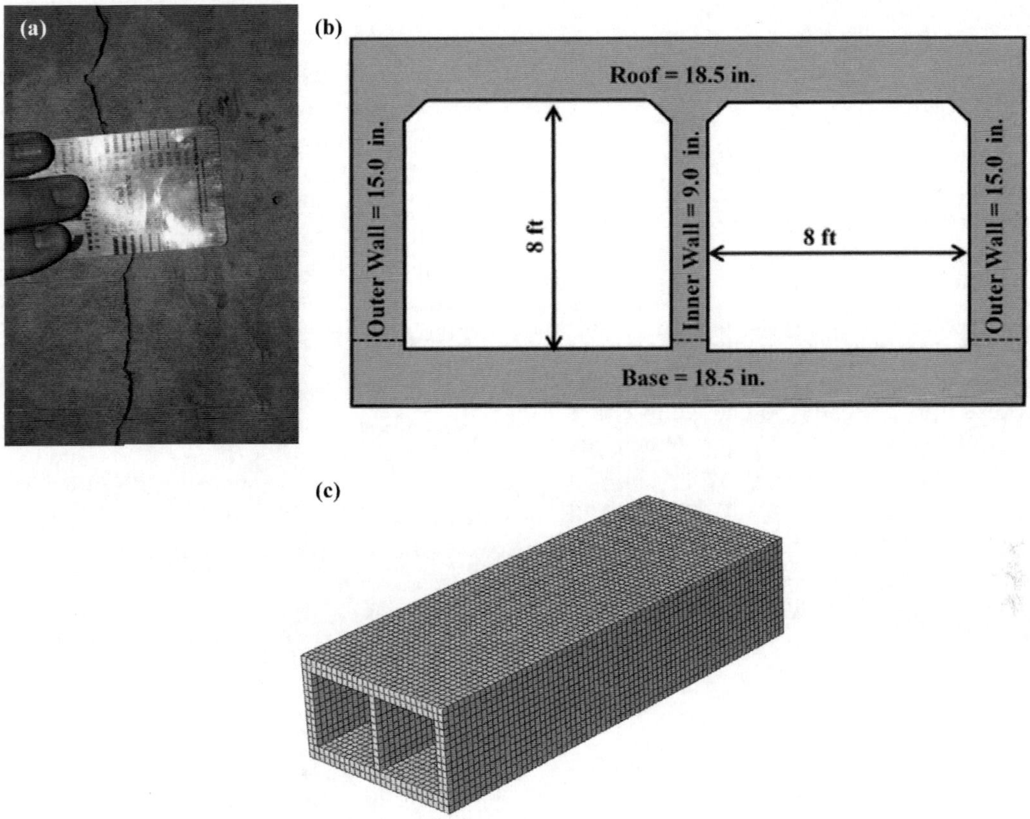

Figure 1. Culvert J information: (*a*) cracking in culvert wall; (*b*) culvert geometry (note that 1 in. = 25.4 mm, 1 ft = 0.305 m); (*c*) finite-element model of one 50 ft culvert section.

achieve this objective, a secondary objective is to develop a representative finite-element model (FEM) that can accurately simulate the early-age stress development of concrete considering the changing mechanical properties, thermal effect, creep or relaxation, and drying shrinkage. Experimental results from restraint to volume change tests with rigid cracking frames were used to verify the accuracy of the FEM. In order to determine measures to mitigate culvert cracking, a parametric study was performed with the FEM to evaluate the effect of changing joint spacing, joint type, construction sequence, concrete coefficient of thermal expansion, placement season, and concrete type.

Causes of Early-Age Cracking

Thermal Effects

The development of thermal stress can be determined with (*3*)

$$\sigma = K_r \times CTE \times \Delta T \times E_c \quad (1)$$

where

σ is the thermal stress (psi),

E_c is the creep-adjusted modulus of elasticity of the concrete (psi),

K_r is the internal/external restraint factor,

CTE is the coefficient of thermal expansion of concrete (in./in./°F),

ΔT is the difference in temperature $= T_{zero\text{-}stress} - T_{min}$ (°F),

$T_{zero\text{-}stress}$ is the temperature at zero stress in the concrete (°F), and

T_{min} is the minimum temperature recorded by the concrete member (°F).

Shrinkage Effects

The loss of moisture from the concrete's cement paste phase due to differences in relative humidity causes drying shrinkage (*4*). Best practice is to cure concrete for at least 7 days after placement, which delays the onset of drying shrinkage. Without curing, the effects of drying shrinkage will be additive to the stresses caused by thermal effects. The shrinkage model from Bažant and Baweja (*5*) is used in this study:

$$\varepsilon_{sh}(t, t_0) = -\varepsilon_{sh\infty} k_h S(t) \quad (2)$$

where

$\varepsilon_{sh}(t,t_0)$ is the mean shrinkage in the cross section at time t related to curing time t_0 (in./in.),

$\varepsilon_{sh\infty}$ is the ultimate drying shrinkage (in./in.),

k_h is a humidity dependent factor (unitless), and

$S(t)$ is the time function defining the shape of the shrinkage curve (unitless).

Autogenous shrinkage is defined as the "change in volume due to chemical process of hydration of cement exclusive of effects of applied load and change in either thermal condition or moisture content" (*6*). When the water-to-cementitious materials ratio (w/cm) is above 0.42, the concrete has enough water for hydration, and the stresses related to autogenous shrinkage can be neglected (*7*). Since all the CIP culverts surveyed in this study had a w/cm of 0.44, the effect of autogenous shrinkage was neglected.

Creep Effects

Creep is the time-dependent increase in strain when concrete is subjected to sustained stress. Often the time-dependent response of concrete is expressed in terms of compliance, which includes both elastic and time-dependent deformations (i.e. creep effects) (*8*). There are many compliance models for hardened concrete in the literature and the Modified B3 Model (*9*) is used in this study. The original B3 Model is based on solidification theory and the Kelvin chain model (*5*). The aging of concrete is modeled as a growth of the volume fraction of load-bearing solidified matter (hydrated cement), which itself is treated as a non-aging viscoelastic material (*10*). The compliance function of the B3 Model can be determined as follows (*5*):

$$J(t, t') = q_1 + C_0(t, t') + C_d(t, t', t_0) \quad (3)$$

$$C_0(t, t') = q_2 Q(t, t') + q_3 ln\ [1 + (t - t')^n] + q_4\left(\frac{t}{t'}\right) \quad (4)$$

where

q_1 is the 28-day concrete modulus of elasticity divided by 0.6 (1/psi),

q_2, q_3 and q_4 are based on concrete compressive strength and proportions (1/psi),

t, t', t_0 are age of concrete, age at loading, and age when drying begins, respectively (days),

$C_0(t,t')$ is the basic creep term (1/psi),

$C_d(t,t',t_0)$ is the drying creep term (1/psi),

$Q(t,t_0)$ is a function of t and t_0 (unitless),

$q_2Q(t,t')$ is the aging viscoelastic compliance (1/psi),

$q_3 ln(1 + (t-t')^n$ is non-aging viscoelastic compliance (1/psi), and

$q_4 ln(t/t')$ is flow compliance (1/psi).

The original B3 Model was primarily developed for hardened concrete, whereas early-age concrete exhibits both rapidly changing elastic properties and high viscoelastic properties (*9*). Østergaard et al. (*11*) provide a modification to q_2 of the B3 Model shown in Equation 5 to capture the high early-age viscoelastic behavior, and Wei and Hansen (*12*) also verify the need for this correction with experimental results. Byard and Schindler (*9*) developed the Modified B3 Model to include

$$q_2' = q_2\left(\frac{t'}{t' - q_5}\right) \quad (5)$$

and a modification shown in the following equation to the early-age elastic compliance, which leaves the later-age elastic compliance from the original B3 Model unchanged:

$$q_1' = q_1\left(\frac{t'}{t' - q_6}\right) = \frac{0.6}{E_{c,28}}\left(\frac{t'}{t' - q_6}\right) \quad (6)$$

where

$E_{c,28}$ is the 28-day modulus of elasticity (psi),

q_5 is the structural setting time (days), and

q_6 is the factor for early-age elastic behavior (days).

Effect of Reinforcement

Sule and Breugel (*13*) performed experiments to study the effect of reinforcement on early-age cracking due to autogenous shrinkage and thermal effects. Their results show that early-age concrete and reinforcement bars experienced similar deformations caused by the heat of hydration due to their similar CTE; therefore, before cracking the reinforcement does not provide restraint to the concrete (*13*). Based on these findings and considering the objectives of this paper, the effect of reinforcement was neglected in the FEM to model early-age cracking.

Finite-Element Modeling of Creep and Shrinkage

Concrete creep and drying shrinkage were modeled with user-defined subroutines in ABAQUS. The programming language Fortran was used to code the Modified B3 Model and shrinkage model into ABAQUS subroutines. The effects of varying in-place concrete temperatures on creep, shrinkage, setting, tensile strength, and modulus of elasticity were accounted for by using the equivalent age maturity method (*14*).

Incorporating Creep Effects in the Finite-Element Model

The ABAQUS material subroutine UMAT was used to define the mechanical constitutive behavior of the concrete. Yu et al. (*15*) cover the theories of the development of the UMAT code to implement the B3 Model in ABAQUS. Changes to the UMAT code were made to obtain accurate predictions of early-age concrete stresses. This approach uses a step-by-step analysis, and in each time step, each integration point of each finite element is considered within the UMAT subroutine. The inelastic creep problem for the structure is reduced to a sequence of elasticity problems by converting the incremental stress–strain relation for each time step, Δt, to a quasi-elastic stress–strain relation as shown in (*15*)

$$\Delta\sigma = E''(\Delta\varepsilon - \Delta\varepsilon'') \quad (7)$$

where

$\Delta\sigma$ is the stress increment during each time step (psi),
E'' is the incremental modulus (psi),
$\Delta\varepsilon''$ is the inelastic strain increments from creep (in./in.), and
$\Delta\varepsilon$ is strain increment from shrinkage and thermal dilation (in./in.).

Equation 7 makes it possible to calculate the strain response for any stress history (creep), and the stress response for any given strain history (relaxation) (*16*).

Incorporating Thermal and Drying Shrinkage Effects in the Finite-Element Model

The user-subroutine UEXPAN was used to calculate incremental thermal strains as a function of concrete temperature (*17*). The concrete CTE was used to calculate the thermal strain for each time step based on the concrete temperature change, ΔT, that occurred during the time step. The drying shrinkage model was also coded into this subroutine considering the concrete properties. The total strain change, $\Delta\varepsilon$, due to thermal and shrinkage effects was then obtained and used to determine the stress increment with Equation 7.

Experimental Work

Test Setup

Restraint to volume change tests with rigid cracking frames (RCF) were used to measure the early-age stress development in various concretes. The RCF consists of a dog-bone shaped formwork with dimensions of $6 \times 6 \times 49$ in. ($150 \times 150 \times 1{,}300$ mm) and a restraining frame in which fresh concrete is placed (*18*). The formwork is lined with plastic and then sealed after concrete placement; therefore, no drying shrinkage develops. The concrete curing temperature in the setup is controlled to simulate typical concrete elements such as bridge decks, mass concrete, concrete culverts, etc. Because of the volume change effects (thermal and autogenous shrinkage), the concrete in the RCF attempts to shrink and expand; however, these movements are restrained by stiff Invar bars that connect the two dog-bone shaped ends of the frame (*18*). Strain gauges on the Invar bars allow the measurement of tension and compression that develops in the restrained concrete specimen.

Prediction of Measured Concrete Stresses

Seventy-two concrete mixtures were tested to assess the early-age stress development by using the RCF. In ABAQUS, the stress development in the concrete specimen was modeled considering the measured concrete temperature profile and a one-hour analysis time step. As examples, two concretes from Byard (*19*) are presented in Figure 2*a* and 2*b*. One mixture (0.42 LS Fall) contained limestone coarse aggregate and the other mixture (0.42 SLW Fall) was a sand-lightweight concrete which contained lightweight coarse aggregate. Both concretes had river sand as fine aggregate, a water-to-cement ratio (w/c) of 0.42, and were placed under fall temperature conditions. In these two examples, the predicted stress development compares very well with the experimental results. Figure 2*c* presents the results for all 72 concretes tested, with a total of 13,356 data points from the RCF modeled by the FEM. The goodness of fit of the prediction obtained from the FEM was evaluated by determining the coefficient of determination (r^2) for all the data collected. Based on the high r^2 value of 0.88, 88% of the experimental variation in early-age concrete stress development is explained by the FEM. The FEM thus provides accurate predictions of the early-age concrete behavior and was used to evaluate the causes of cracking in CIP concrete box culverts.

Finite-Element Culvert Model

Culvert J, with geometry as shown in Figure 1*b*, is one of the culverts that exhibited extensive cracking on the AEB Project and was used as baseline for the FEM. It has a total length of 892 ft (271 m) with tied joints (with continuous longitudinal reinforcement) every 50 ft (15 m) (*1*). Eight-node linear hexahedral (brick) elements with three translational degrees of freedom at each node (C3D8) were used. The three-dimensionally meshed FEM of one 50 ft section of Culvert J is shown in Figure 1*c*. The finite-element size used for this model is approximately 8 in. (203 mm). At the start of this project, a 1 in. (25 mm) mesh size model was analyzed and these stress results were within 1% when compared with those of the 8 in.

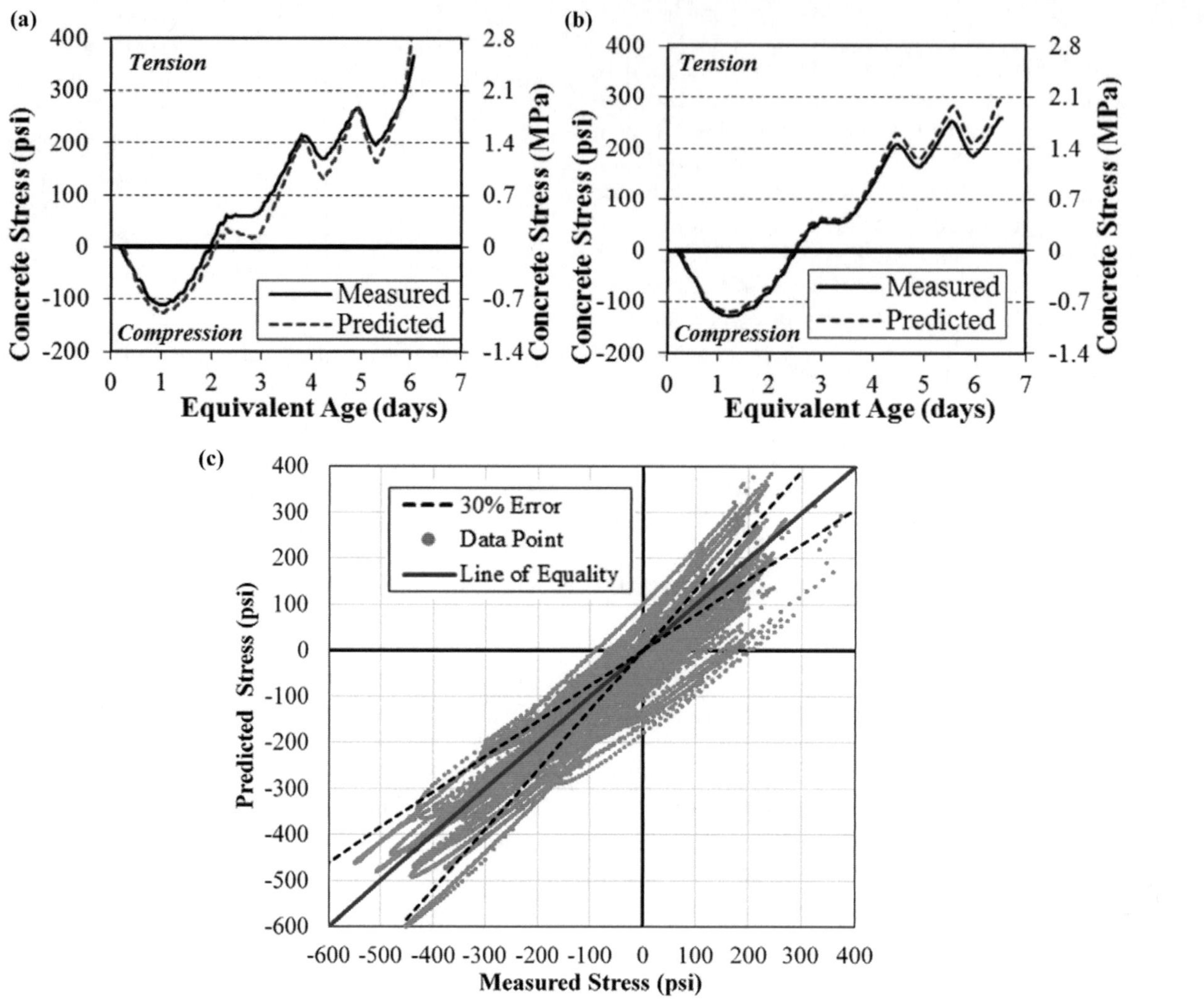

Figure 2. Calculated and measured stress development in rigid cracking frame: (*a*) *w/c* = 0.42 (limestone), (*b*) *w/c* = 0.42 (sand-lightweight concrete), (*c*) all data points.

model. The effects of gravity of all culvert components was included in the model, as this affects the interaction with its supporting subbase. The FEM was used to perform parametric studies to determine the influence of changing the contraction joint spacing, two concrete CTE values, three placement seasons, two joint types, three construction sequences, and three concrete types on early-age cracking risk.

Material Properties

The mixture proportions and other parameters of the concrete used for Culvert J are presented in Table 1. An average environmental humidity of 70% and start of exposure to drying at an age of 7 days was used for the drying shrinkage model.

Concrete Temperature Profile

The concrete temperature profile used for the stress calculation in the FEM was obtained from ConcreteWorks, which is an early-age concrete temperature development and thermal stress analysis software (*20*). The software is capable of analyzing environmental, construction, and concrete mixture proportioning parameters in various types of concrete elements (*20*). The concrete temperature profiles for summer, fall, and winter placement conditions were determined for Montgomery, Alabama, on construction dates of July 30, October 15, and December 15, respectively.

Staged Construction of Culvert Components

Since the base is cast first and the wall and roof are cast later, and more culvert sections are also later constructed,

Table 1. Concrete Properties of Culvert J

Parameter	Value
Water	275 lb/yd^3 (163 kg/m^3)
Type I/II Portland Cement	465 lb/yd^3 (276 kg/m^3)
Class C Fly Ash	155 lb/yd^3 (92 kg/m^3)
No.67 Limestone	1,857 lb/yd^3 (1,102 kg/m^3)
Fine Aggregate	1,283 lb/yd^3 (761 kg/m^3)
Total Air Content	4–6%
w/cm	0.44
28-day Compressive Strength	4,000 psi (27.6 MPa)
28-day Modulus of Elasticity	3,600 ksi (24.8 GPa)
Activation Energy	40 kJ/mol
Coefficient of Thermal Expansion	5.52×10^{-6}/°F (9.94×10^{-6}/°C)
Initial Set (hours)	5.0

the parts of the culvert concrete have different ages and thus different mechanical properties, creep, shrinkage, and temperature effects. The whole model was divided into several analysis steps to account for each construction stage. The sequential stages used to construct a culvert were simulated by initially creating all elements of the FEM; however, the elements in later stages were deactivated at first and then progressively activated and their age corrected as needed to simulate when they were cast. A tie constraint (*21*) was used to connect previously cast concrete surfaces to newly cast concrete surfaces, except where a contraction joint was used to separate the movements between sections.

Modeling of Ground Restraint

Culverts in Alabama are typically constructed on a crushed-stone subbase, and spring elements were used to model the behavior at this interface. Several studies on the frictional behavior of concrete pavement and subbase show the relationship is nonlinear with an elastic stiffness followed by a plastic response. Wesevich et al. (*22*) report an elastic stiffness value of 220 psi/in. (60 MPa/m) for concrete on ¾ in. (19 mm) crushed stone. Jeong et al. (*23*) performed push-off field test of a concrete slab cast on a crushed-stone subbase and reported an elastic stiffness of frictional resistance of approximately 507 psi/in. (138 MPa/m) which was used in this FEM. Because of the complexity of nonlinear hysteresis behavior between the base and subgrade, a linear spring was used for the *horizontal* connection to simulate the frictional resistance. For the *vertical* connection between concrete base and subgrade, a constant spring stiffness was used when the spring elements experience compression, and zero spring stiffness was assigned (no tension resisted) when the culvert base lifted up from the subbase.

Assessment of Cracking Risk

The cracking failure criterion for the model was based on the tensile stress-to-strength ratio as presented by Riding et al. (*24*). The cracking risk classification is based on probability density function, which was obtained from the distribution of the tensile stress-to-strength at cracking from experimental data. When the tensile stress-to-strength ratio is above 0.67, the cracking probability was 75%, and at this ratio the cracking risk was defined as being at a "high" level (*24*). Kanstad et al. (*25*) also used FEM and found that when the tensile stress-to-strength ratio was in the range of 0.75 to 0.88, all walls of the 1,115 ft (340 m) long Maridal culvert in Norway showed one to four cracks in each 50 ft (15 m) long section.

Results and Discussions

Culvert J Model Results

Figure 3*a* shows the temperature history of the culvert concrete and the ambient temperature used for the FEM. After concrete placement, the temperature starts to rise because of the co-occurrence of heat of hydration and solar radiation peak effects. After reaching the peak temperature, the concrete gradually cools down until it is primarily influenced by solar and ambient temperature cycles.

The longitudinal stress development in the culvert base, wall, and roof compared with the tensile strength development with time is shown in Figure 3*b*. Negative values represent compressive stresses and positive represent tensile stresses. The wall and roof are cast 7 days later than the base. All FEM stress results are from mid-length of the 50 ft (15 m) section, which is where the restraint is the highest. For the culvert *wall*, results revealed that the maximum stress occurs near their interface with the base, which is where wall stresses are plotted in Figure 3*b*. Immediately following final set, compressive stresses develop due to the rise in concrete temperature. When the concrete begins to cool, tensile stresses develop because of the restraint of concrete contraction. Tensile stresses in the roof are much lower than in the base or wall, because the restraint to movement provided by the base is much lower at the level of the roof.

Based on the stress and strength development shown in Figure 3*b*, the tensile stress-to-strength ratio over time was determined and is graphed for the base, wall, and roof in Figure 3*c*. From this figure, the cracking risk can

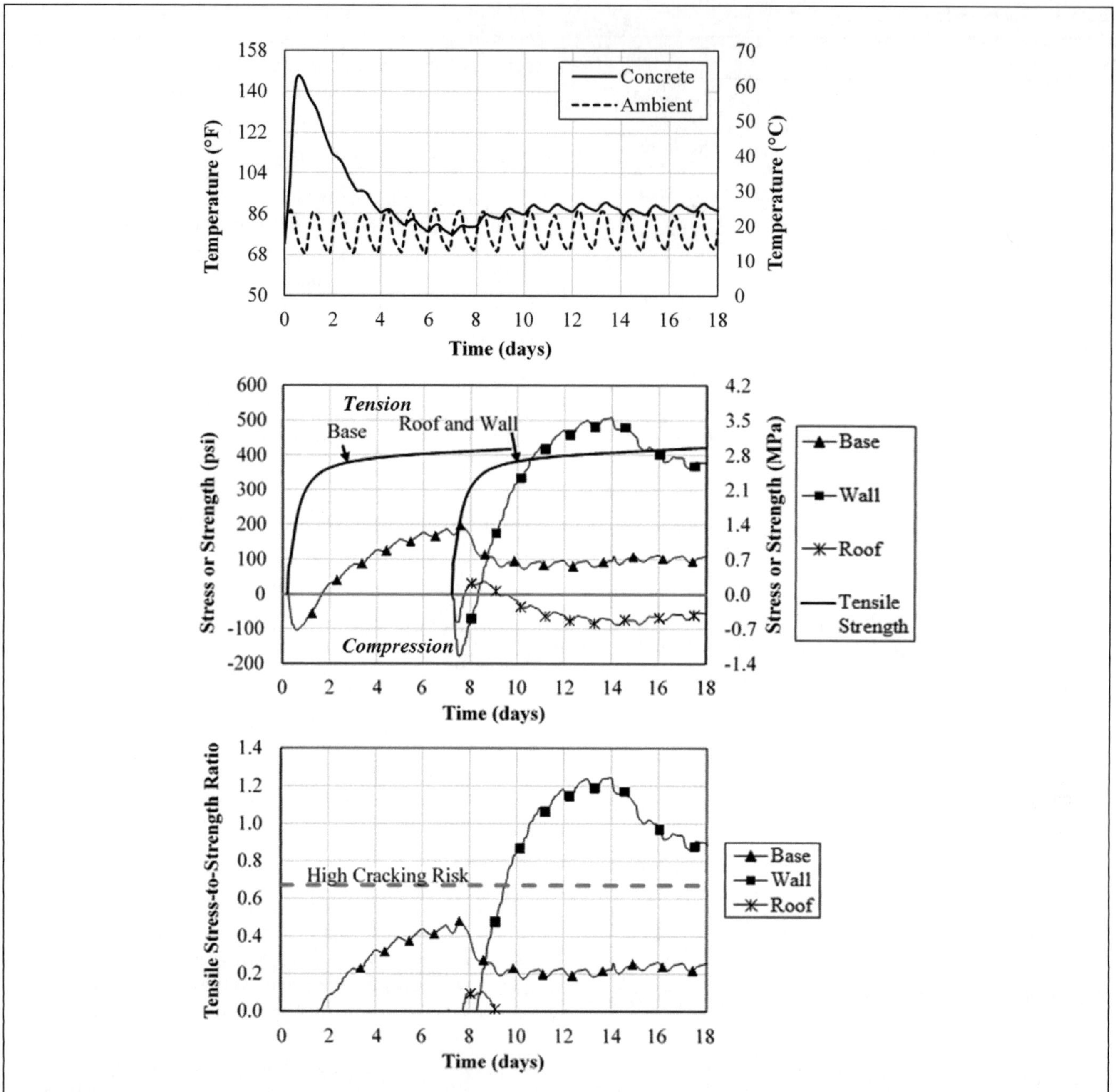

Figure 3. Summer placement results for one section of Culvert J versus time: (*a*) temperatures, (*b*) stress and strength, (*c*) tensile stress-to-strength ratio.

be assessed, and a dashed line at 0.67 above which the cracking risk is considered "high" (*24*) is shown for reference. It is clear from Figure 3*c* that the cracking risk in the wall quickly develops above the high threshold and remains high for many days, whereas the cracking risk in the base is elevated but remains below the high cracking risk threshold, and the cracking risk in the roof is very low. This finding explains why many of the surveyed culverts had wide cracks primarily in the walls and bases.

Effect of Contraction Joint Spacing

The effect of contraction joint spacing, assuming each length of culvert section is free to expand or contract from each other, on the cracking risk of culvert is evaluated here. The following six contraction joint spacings were evaluated: 18 ft, 24 ft, 30 ft, 36 ft, 42 ft, and 50 ft (5.5 m, 7.3 m, 9.1 m, 11 m, 13 m, and 15 m). In this analysis, the base was cast first and wall and roof were added 7

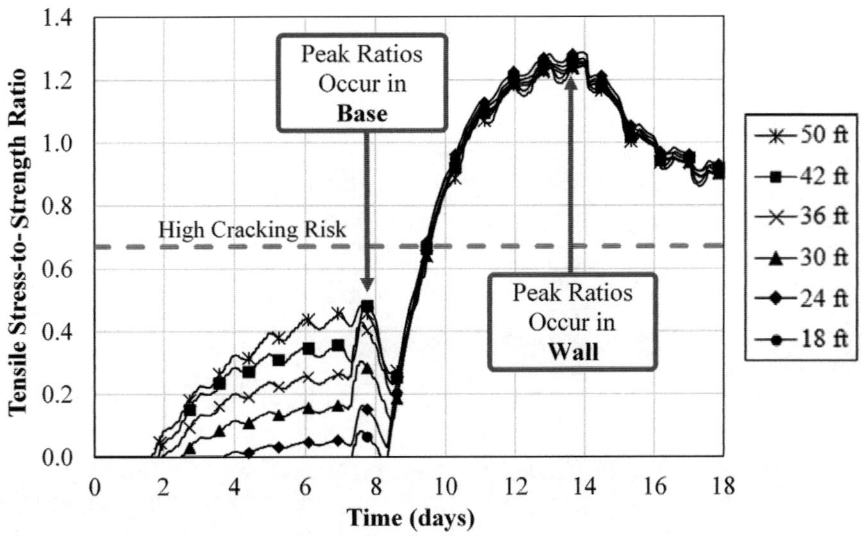

Figure 4. Effect of culvert contraction joint spacing on culvert cracking risk.

days later. The maximum tensile stress-to-strength ratio *envelope* for the six contraction joint spacings is presented in Figure 4. From 0 to 8 days, the cracking risk of the *base* significantly decreases with decrease in contraction joint spacing. However, from 8 days and later it can be seen that the cracking risk in the culvert *wall* changes little (less than 10%) with decrease in contraction joint spacing. As will be shown in the rest of this section, this is because the wall is highly restrained against movement by the base on which it is cast, so a change in contraction joint spacing does not significantly impact the wall stresses. This finding is similar to that of Bosnjak (*26*) who concluded that short and long culvert wall sections have similar maximum stresses.

Longitudinal stress contour plots of the culvert wall for short and long contraction joint spacings of 18 ft (5.5 m) and 42 ft (13 m) are shown in Figure 5*a*. The maximum tensile stress of approximately 500 psi (3.45 MPa) occurs near the bottom of the wall for both small and large contraction joint spacings. The stress decreases from the bottom to the top of the wall, and from mid-length to the end of the wall because this matches the change in restraint to movement provided. Based on the stress contour plot and location of maximum stresses, possible cracks in the short and long culvert wall are shown in Figure 5*b* and 5*c*.

Effect of Joint Type and Construction Sequence

An elevation view of Construction Sequence A is shown in Figure 6*a*, which is similar to the actual construction sequence used to construct Culvert J. It includes three 50 ft (15 m) culvert sections, consisting of three walls and roofs cast on three bases. The ages in Figure 6*a* represent the days when the culvert component is constructed relative to the start of construction of the first base. With this construction sequence, the delay between casting adjacent base sections is 2 days, the delay between adjacent wall sections is 2 days, and the delay between starting the first base and wall sections is 7 days.

The effect of joint type when using Construction Sequence A on the stress-to-strength ratio in the culvert base and wall is shown in Figure 6*b*. From Figure 6*b* it can be concluded that the joint type significantly influences the cracking risk of the *base*; however, the joint type does not have a significant impact on the cracking risk of the *wall*. The maximum cracking risk in the base with a tied joint decreases from 1.23 to 0.72 when a contraction joint is used. Since the wall is highly restrained against movement by the base on which it is cast, a change in joint type has a small impact on the *wall* stresses.

Two additional construction sequences (B and C) were evaluated to determine if a change in construction sequence will reduce the occurrence of cracking. In Construction Sequences B and C, the time delay between base sections is 2 days and 4 days, the delay between wall sections is 6 days and 4 days, and the delay between starting the first base and wall sections is 4 days and 2 days respectively. For Construction Sequences A, B, and C, the delay between casting the wall and base is 7 days, 4 days, and 2 days, respectively.

The maximum tensile stress-to-strength ratio in the base and wall for the three construction sequences modeled is shown in Figure 6*c*. In this figure, the first six columns for the culvert base show a decreasing trend of

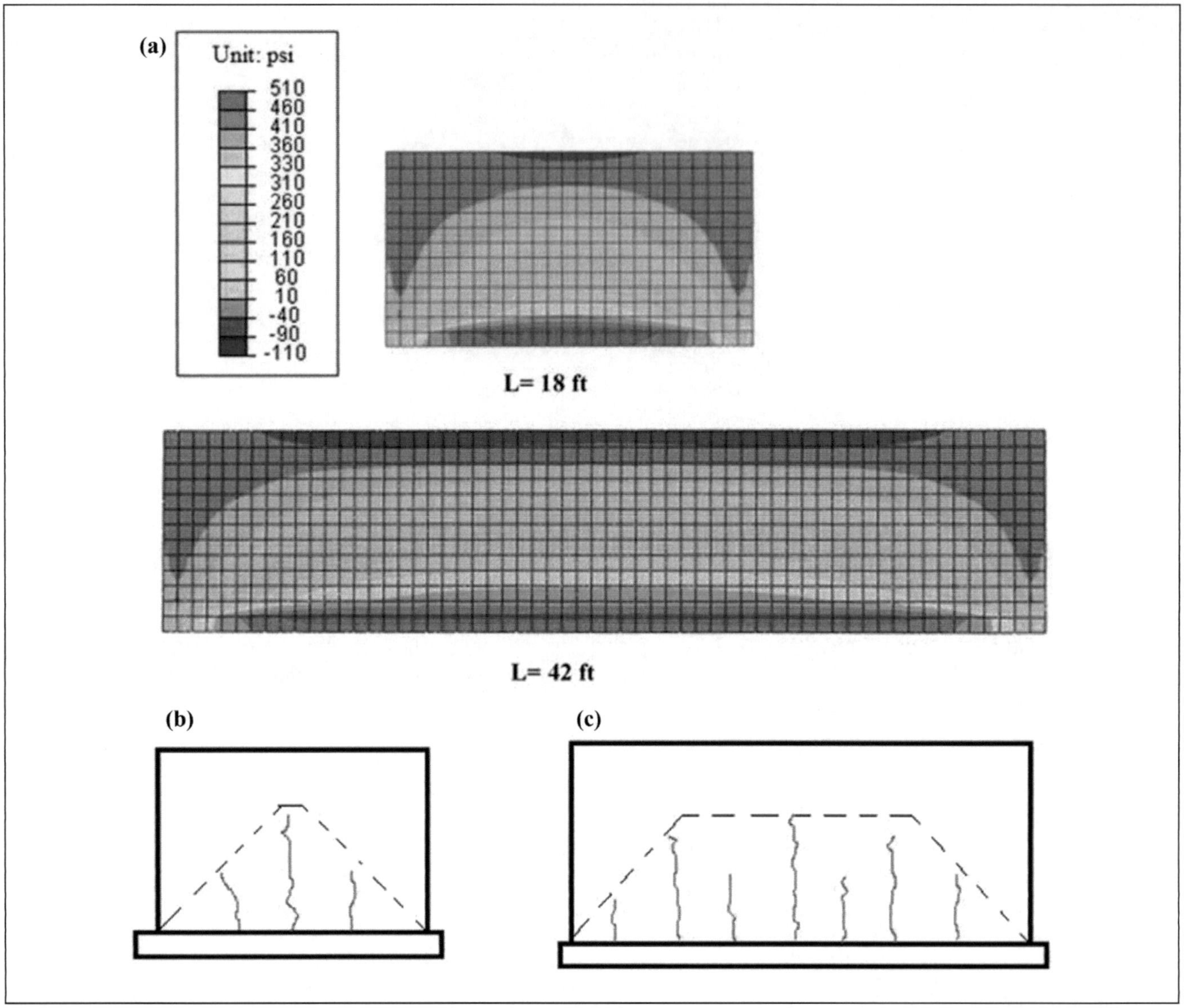

Figure 5. Impact of contraction joint spacing on wall stresses: (*a*) longitudinal stress contours for short and long contraction joint spacings (Note: 1 MPa = 145 psi), (*b*) possible cracks for L = 18 ft, (*c*) possible cracks for L = 42 ft. [(*b*) and (*c*) adapted from (*27*)].

maximum cracking risk for both tied joints (grey columns) and contraction joints (white columns). Similarly, the cracking risk in the culvert wall also decreases systematically when changing from Construction Sequence A to B to C. From these results, it can be concluded that the younger the base when the wall is cast, the lower the stresses that develop in the *wall*. This is because the younger the base when the wall is cast on it, the lower the stiffness of the base, which reduces the restraint that the base provides for the wall.

Effect of Miscellaneous Variables

Placement Season. The effect of constructing culverts during different seasons on the cracking risk is evaluated in this section. Figure 7*a* shows the maximum stress-to-strength envelope for culverts constructed under summer, fall, and winter conditions. Summer placement leads to a much higher cracking risk, while winter placement leads to a much lower cracking risk, with fall placement in between these two extremes. This means that extensive culvert cracking is much more likely when construction occurs during hot weather conditions.

Concrete Coefficient of Thermal Expansion. The coarse aggregate type significantly affects the concrete's coefficient of thermal expansion (CTE) (*28*). The stresses in culverts made with river gravel concrete with CTE of 6.95×10^{-6}/°F (12.51×10^{-6}/°C) and limestone concrete with CTE of 5.52×10^{-6}/°F (9.93×10^{-6}/°C) were evaluated (*28*). The results in Figure 7*b* indicate that the use

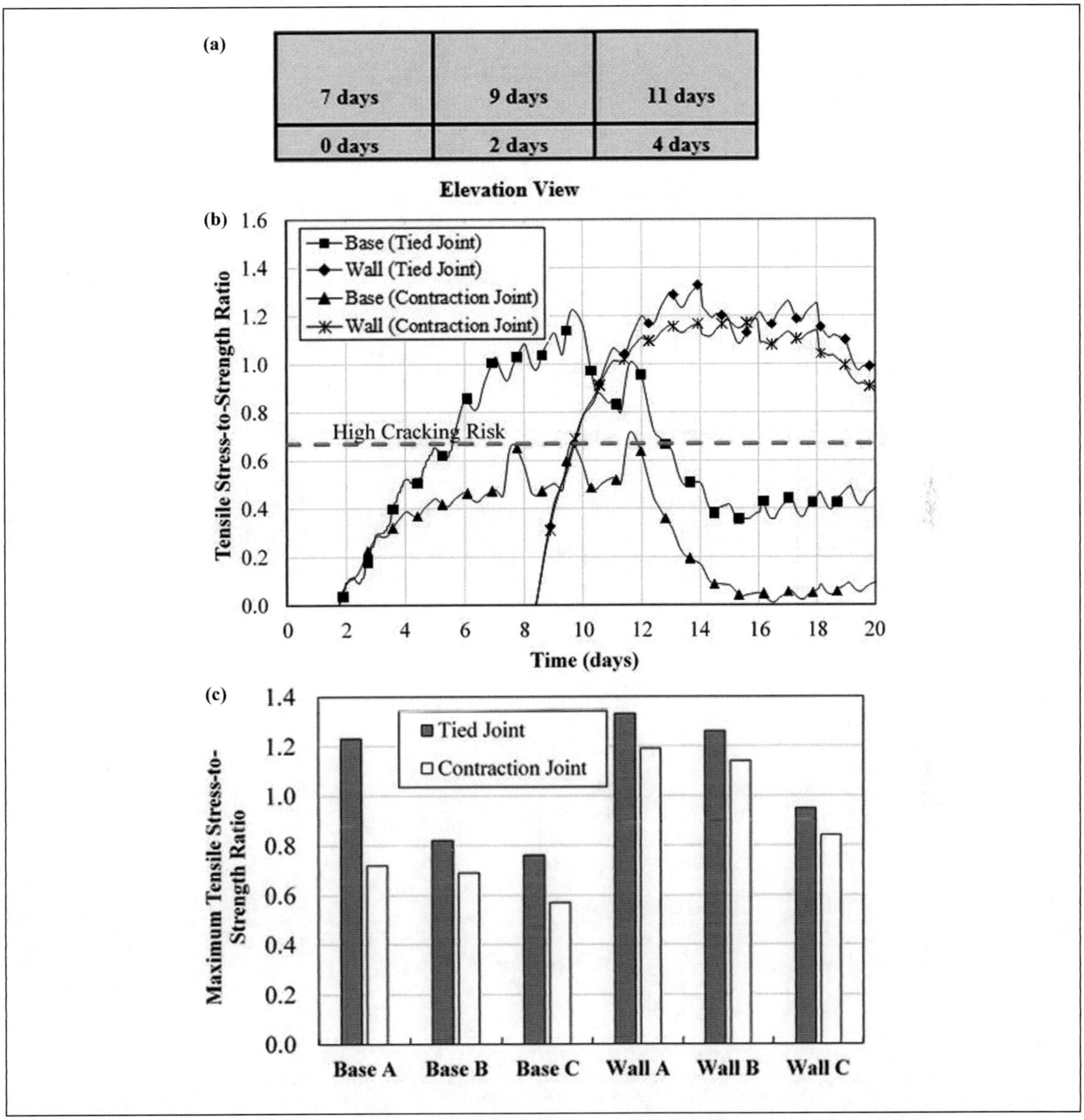

Figure 6. Effect of construction sequence and joint type on culvert cracking risk: (*a*) elevation view of Construction Sequence A, (*b*) cracking risk for two joint types, (*c*) cracking risk for three construction sequences.

of a lower CTE concrete will decrease the cracking risk in the base and wall; however, due to the high restraint, the stresses in the wall remain above the high cracking risk threshold.

Use of Lightweight Aggregate Concrete. During the construction sequence analysis it was shown that the lower the stiffness of the base, the lower the stresses that develop in the wall. Based on this finding, the effect of adding lightweight aggregate to the concrete to reduce its modulus of elasticity was analyzed. Sand-lightweight (SLW) concrete and all-lightweight (ALW) concrete were used to evaluate the cracking risk compared with using normal-weight (NW) concrete. The properties of SLW and ALW concrete were obtained from Byard (*19*) who tested the stress development of various concretes made with lightweight

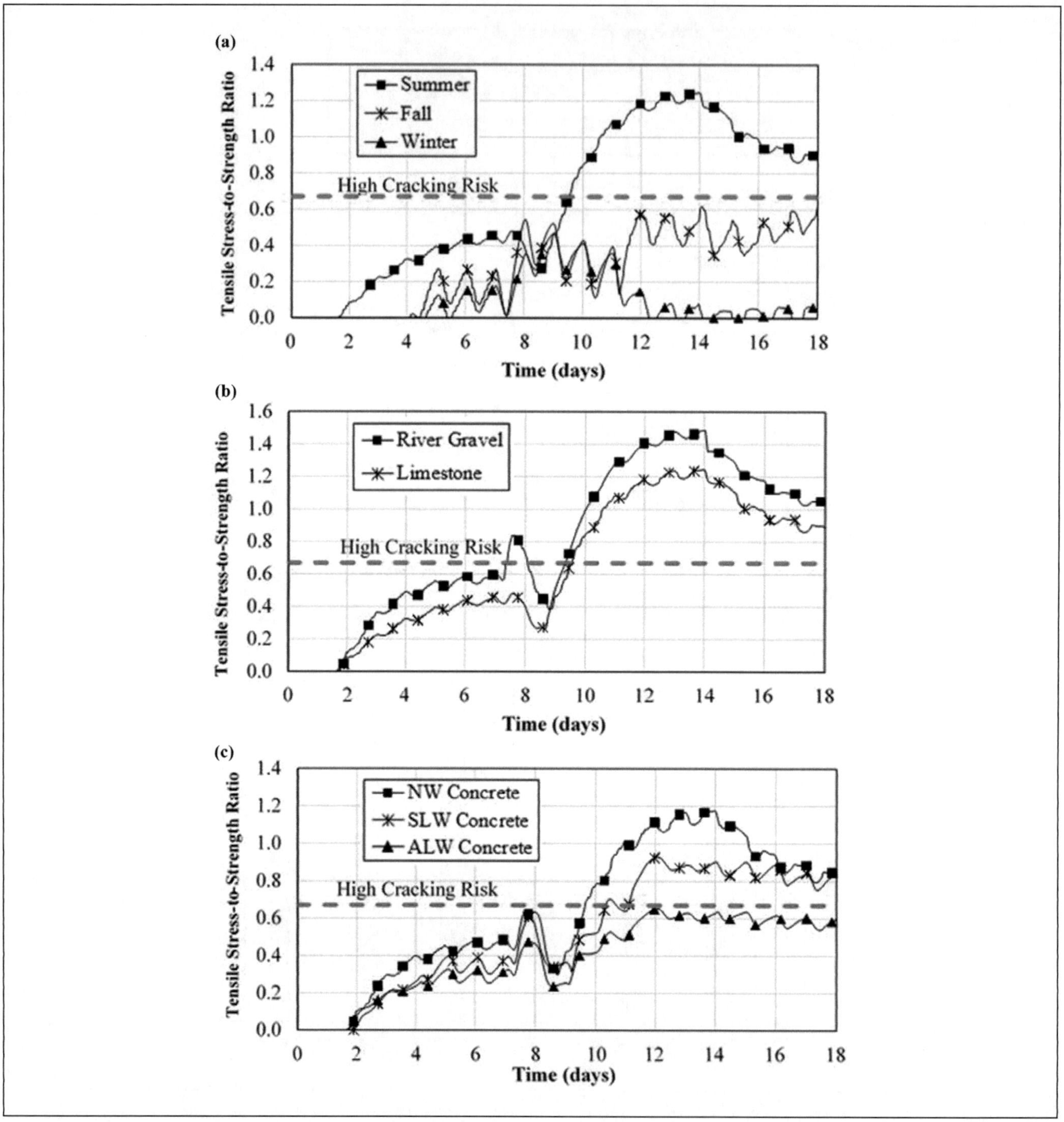

Figure 7. Effect of miscellaneous variables on cracking risk: (*a*) placement season, (*b*) concrete CTE, (*c*) concrete types.

aggregates. The SLW and ALW concretes have a *w/c* of 0.42 and their temperature profiles were obtained from ConcreteWorks using their mixture properties. Figure 7*c* reveals that use of ALW and SLW concrete significantly decrease the cracking risk in the culvert compared with using NW concrete. The ALW has the lowest cracking risk among the three concrete types, and this is attributed to the 34% reduction in modulus of elasticity when using ALW as compared with NW concrete. This finding is similar to that of Byard (19) who concluded that ALW and SLW can decrease early-age cracking of bridge deck concrete.

Conclusions

A three-dimensional finite-element model (FEM) of CIP concrete culverts was developed. The FEM of early-age stress development of concrete considering temperature effects, creep, and drying shrinkage was verified by experimental results. The calibrated FEM was used to determine the influence of changing the contraction joint spacing, two concrete CTE values, three placement seasons, two joint types, three construction sequences, and three concrete types on early-age cracking risk. Results from this study support the following conclusions:

1. The cracking risk of the culvert wall, base, and roof was found to be high, moderate, and low, respectively, which explains why many surveyed culverts had wide cracks primarily in walls and bases.
2. Contraction joint *spacing* only significantly influences the cracking risk in the culvert *base*, but has little effect on the cracking risk of the *wall*. This is because the base provides a very high degree of restraint to the wall.
3. Use of contraction joints can help significantly decrease the cracking risk in culvert *base* compared with use of tied-construction joints, but they do not have a significant effect on the maximum cracking risk in the culvert *wall*.
4. Use of a construction sequence that attains the youngest (lowest stiffness) *base* when the wall is cast reduces the cracking risk in the *wall*.
5. Hot weather construction leads to the highest cracking risk, while cold weather placement results in a much lower cracking risk, with fall placement conditions in between these two extremes.
6. Use of concrete with a lower coefficient of thermal expansion decreases the cracking risk in the base and wall.
7. Use of sand-lightweight concrete and all-lightweight concrete significantly decrease the cracking risk in the culvert compared with using normal-weight concrete. The all-lightweight concrete most decreases the cracking risk among the three concrete types evaluated.

Acknowledgments

The financial support provided by the China Scholarship Council is appreciated. The authors would also like to thank Qiang Yu of the University of Pittsburgh for providing an initial version of the UMAT code. The support of the Alabama Department of Transportation is also appreciated.

Author Contributions

The authors confirm contribution to the paper as follows: study conception and design: Liu, Schindler; data collection: Liu, Schindler; analysis and interpretation of results: Liu, Schindler, Davidson; draft manuscript preparation: Liu, Schindler, Davidson. All authors reviewed the results and approved the final version of the manuscript.

References

1. Minton, L. W. *Mitigation of Cracking in Cast-in-Place Reinforced Concrete Box Culverts in Alabama.* MS thesis. Auburn University, Auburn, Ala., 2012.
2. ACI Committee 224. *Report on Control of Cracking in Concrete Structures.* American Concrete Institute, Farmington Hills, Mich., 2001.
3. Schindler, A. K., and B. F. McCullough. The Importance of Concrete Temperature Control during Concrete Pavement Construction in Hot Weather Conditions. *Transportation Research Record: Journal of the Transportation Research Board*, 2002. 1813: 3–10.
4. Mehta, P. K., and P. J. Monteiro. *Concrete: Microstructure, Properties, and Materials.* McGraw-Hill Publishing, New York, NY, 2006.
5. Bažant, Z. P., and S. Baweja. Creep and Shrinkage Prediction Model for Analysis and Design of Concrete Structures: Model B3. *ACI Special Publications*, Vol. 194, 2000, pp. 1–84.
6. ACI Committee CT. *ACI Concrete Terminology.* American Concrete Institute, Farmington Hills, Mich., 2016.
7. Mindness, S., J. Young, and D. Darwin. *Concrete*, 2nd ed. Prentice Hall, Upper Saddle River, N.J., 2002.
8. ACI Committee 209. *Report on Prediction of Creep, Shrinkage, and Temperature Effects in Concrete Structures.* American Concrete Institute, Farmington Hills, Mich., 1992.
9. Byard, B. E., and A. K. Schindler. Modeling Early-Age Stress Development of Restrained Concrete. *Materials and Structures*, Vol. 48, No. 1-2, 2015, pp. 435–450.
10. Bažant, Z. P., and S. Prasannan. Solidification Theory for Concrete Creep. I: Formulation. *Journal of Engineering Mechanics*, Vol. 115, No. 8, 1989, pp. 1691–1703.
11. Østergaard, L., D. A. Lange, S. A. Altoubat, and H. Stang. Tensile Basic Creep of Early-Age Concrete Under Constant Load. *Cement and Concrete Research*, Vol. 31, No. 12, 2001, pp. 1895–1899.
12. Wei, Y., and W. Hansen. Tensile Creep Behavior of Concrete Subject to Constant Restraint at Very Early Ages. *Journal of Materials in Civil Engineering*, Vol. 25, No. 9, 2012, pp. 1277–1284.
13. Sule, M., and K. V. Breugel. The Effect of Reinforcement on Early-Age Cracking due to Autogenous Shrinkage and Thermal Effects. *Cement and Concrete Composites*, Vol. 26, No. 5, 2004, pp. 581–587.
14. Carino, N. J. The Maturity Method. In *Handbook on Nondestructive Testing of Concrete* (Malhotra, V. M., and N. J. Carino, eds.), ASTM International, West Conshohocken, Pa., 2004, pp. 5.1– 5.47.

15. Yu, Q., Z. P. Bažant, and R. Wendner. Improved Algorithm for Efficient and Realistic Creep Analysis of Large Creep-Sensitive Concrete Structures. *ACI Structural Journal*, Vol. 109, No. 5, 2012, pp. 665–675.
16. Bažant, Z. P., Q. Yu, and G. Li. Excessive Long-Time Deflections of Prestressed Box Girders. II: Numerical Analysis and Lessons Learned. *Journal of Structural Engineering*, Vol. 138, No. 6, 2012, pp. 687–696.
17. Dassault Systèmes Simulia Corp. *Abaqus User Subroutines Reference Manual*. Providence, R.I., 2012.
18. Byard, B. E., A. K. Schindler, R. W. Barnes, and A. Akash. Cracking Tendency of Bridge Deck Concrete. *Transportation Research Record: Journal of the Transportation Research Board*, 2010. 2164: 122–131.
19. Byard, B. E. *Early-Age Behavior of Lightweight Aggregate Concrete*. Doctoral dissertation. Auburn University, Auburn, Ala., 2011.
20. Riding, K. A. *Early Age Concrete Thermal Stress Measurement and Modeling*. Doctoral dissertation. University of Texas, Austin, Tex., 2007.
21. Dassault Systèmes Simulia Corp. *Abaqus/CAE User Manual*. Providence, R.I., 2012.
22. Wesevich, J. W., B. F. McCullough, and N. H. Burns. *Stabilized Subbase Friction Study for Concrete Pavements*. FHWA/TX-87 + 459-1. FHWA, U.S. Department of Transportation, 1987.
23. Jeong, J., J. Park, J. Lim, and S. Kim. Testing and Modelling of Friction Characteristics Between Concrete Slab and Subbase Layers. *Road Materials and Pavement Design*, Vol. 15, No. 1, 2014, pp. 114–130.
24. Riding, K. A., J. L. Poole, A. K. Schindler, M. C. G. Juenger, and K. J. Folliard. Statistical Determination of Cracking Probability for Mass Concrete. *ASCE Journal of Materials in Civil Engineering*, Vol. 26, No. 9, 2014, p. 13.
25. Kanstad, T., P. F. Takács, and D. Bosnjak. Field Measurements versus Calculated Results for Norwegian High Performance Concrete Structures. In *Finite Elements in Civil Engineering Applications: Proceedings of the Third Diana World Conference*, CRC Press, Tokyo, Japan, 9–11 October 2002, p. 107.
26. Bosnjak, D. *Self-Induced Cracking Problems in Hardening Concrete Structures*. Doctoral dissertation. The Norwegian University of Science and Technology, Trondheim, Norway, 2000.
27. RILEM Technical Committee Recommendation 119-TCE. Assessment of External Restraint. In *RILEM Report 15, Prevention of Thermal Cracking in Concrete at Early Ages* (Springenschmid, R., ed.), E&FN Spon, London, 1998, pp. 149–176.
28. Schindler, A. K., M. L. Hughes, R. W. Barnes, and B. E. Byard. *Evaluation of Cracking of the US 331 Bridge Deck*. FHWA/ALDOT 930-645. FHWA, U.S. Department of Transportation, 2010.

The Standing Committee on Basic Research and Emerging Technologies Related to Concrete (AFN10) peer-reviewed this paper (18-00760).

The contents of this paper reflect the views of the authors, who are responsible for the facts and accuracy of the data presented. The contents do not reflect the official views or policies of the Alabama Department of Transportation.

TRR
JOURNAL OF THE TRANSPORTATION RESEARCH BOARD

Article

Transportation Research Record
2018, Vol. 2672(27) 37–44

DOI: 10.1177/0361198118791912
journals.sagepub.com/home/trr
SAGE

Mechanical Analysis of Unconfined Compression Test of Cylinder

Jian Xiao[1], Zhiming Tan[1], and Jingliang Chen[2]

Abstract

The unconfined compression test is widely used because of convenient specimen formation, simple experimental process, and low requirement for equipment. In view of the experimental characteristics, the axisymmetric mechanical model of a cylinder with horizontal friction was constructed and its elastic analytic solution was also deduced. The stress and the strain of the cylinder under unconfined compression with different boundary conditions were analyzed by using the analytical solution, and the failure characteristics of the cylinder under unconfined compression were studied and verified by laboratory testing. The results showed that the end restraint effect of the cylinder increased with the decrease of the height-to-diameter ratio and the increase of the Poisson ratio. When the end surface of the cylinder was constrained, the maximum tensile strain appeared in the middle of the cylindrical side; its value increased with the constraint effect, and greater loads were needed to make the same vertical displacement. The compressive strength derived from the current method would be too weak, and the compressive modulus of elasticity would be too great. If the end constraint was large, a small network crack first formed in the middle of the outer side of the cylinder, forming the double-tapered destruction. If the end face was weakly restrained, the cracks descended from the top to the bottom, and the specimens split.

The unconfined compression test is mainly used to determine the compressive strength and elastic modulus of civil engineering materials (*1–3*). However, during data processing, the constraining action of the rigid plate of the press on the lateral displacement of the pressure face at both ends of the specimen is neglected. Therefore, the compressive strength derived from the current method would be too weak, and the compressive elastic modulus would be too great. Investigating the influence of end restraint on the failure of specimens, various scholars have formed different mechanical models to discuss the above problem. In 1902, Filon adopted polynomials and trigonometric functions to separate the variables in both radial and vertical directions. He discussed the mechanical state of the short cylinder with full restraints at both ends and with the height diameter ratio of $\pi/3$ under conditions of unconfined compression. He indicated that the initial crack position of the specimen was located around the end, and then the rupture extended to form the vertebral body on the end face (*4*). In 1976, Beisetaev and Nikiforovskii simplified the cylinder into a two-dimensional plane, and studied it by the two-dimensional finite element method. The face constraint and Poisson's ratio was found to have a great influence on the stress and strain distribution, but they did not explore the impact of the law in depth (*5*). At this point, the theoretical study of the unconfined compression test was also maintained in the two-dimensional plane solution stage. In 1952, the mechanical problem of a cylinder was solved through the Love displacement function, which was built by using the Bessel function, the virtual Bessel function, the trigonometric function, and the hyperbolic function (*6*), but the influence of end restraint on the failure of specimens was not discussed, and it only provided a method of analysis. In 1992, the Saitou Hideo method (*6*) was used by Watanabe Masahira, and the stress–strain state of the cylinder under uniaxial and triaxial compression conditions was discussed under the premise that the horizontal displacement of the end face was the quadratic function of the diagonal distance (*7*). However, the assumptions of horizontal displacement were slightly flawed and lacked mechanical significance. In 2000, Chau proposed a Love displacement function for a cylinder that can be used to solve under any load (*8*), but

[1]Key Laboratory of Road and Traffic Engineering of Ministry of Education & School of Transportation Engineering, Tongji University, Shanghai, P.R. China
[2]Shanghai Municipal Engineering Design Institute (Group) Co., LTD., Shanghai, P.R. China

Corresponding Author:
Address correspondence to Zhiming Tan: 13901779114@126.com

Chau discussed the case where the end face was locally loaded and the outer side of the cylinder was loaded, and did not study unconfined compression test conditions. Therefore, it was necessary to analyze the unconfined compression problem of the cylinder and find out the influence of the horizontal constraint on the stress–strain state of the specimen, so as to make suggestions for the improvement of the unconfined compression test on cylinders.

In this study, the axisymmetric mechanical model of a cylinder with horizontal friction was constructed and its elastic analytic solution was also deduced. The stress and the strain of the cylinder under unconfined compression with different end boundary conditions were analyzed. The failure characteristics of the cylinder under unconfined compression were studied and verified by laboratory testing.

Elastic Solutions of Unconfined Compression of Cylinders

A homogeneous and isotropic elastic cylinder of radius R and height $2h$ in cylindrical coordinate system is shown in Figure 1. The axial external load is P, and the origin of the cylindrical coordinates is at the center of the cylinder. The radius direction is set as the r axis, the vertical direction is set as the z axis, let $\rho = r/R$, $\psi = z/h$, $\kappa = h/R$, where κ is the height-to-diameter ratio of the cylindrical specimen (Figure 1).

The model is symmetrical, taking half of it for analysis. The radial shear stress $\tau(1, \psi)$ of the upper end of the specimen was assumed to be proportional to its radial displacement $u(1, \psi)$, and its proportional constant is denoted as k_u, which is a horizontal spring friction coefficient. The boundary conditions of the two end faces of a cylinder can be expressed as:

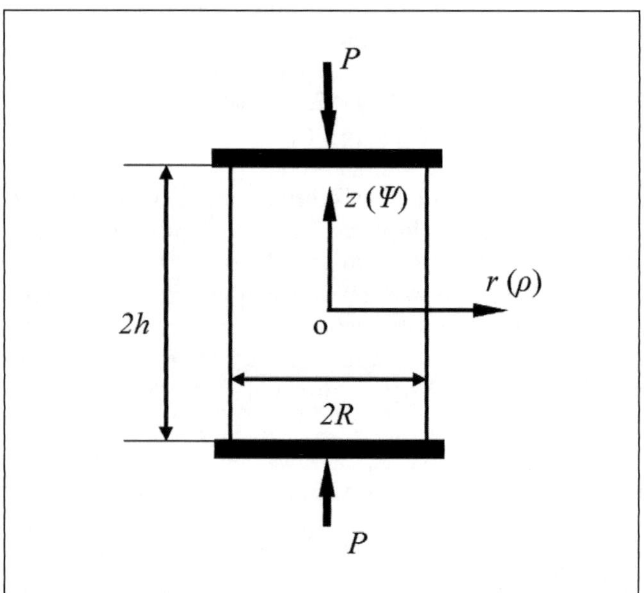

Figure 1. Schematic diagram of unconfined compression test of Cylinder.

$$\left.\begin{aligned} w(\rho, 1) = -w_0, \quad k_\tau \tau_{rz}(\rho, 1) &= -k_u u(\rho, 1) \\ w(\rho, 0) = 0, \quad \tau_{rz}(\rho, 0) &= 0 \end{aligned}\right\} \tag{1}$$

where k_τ is the function of 0, 1, $k_\tau = 0$: the radial displacement of the end face is completely constrained; $k_\tau = 1$: the radial displacement of the end face is partially constrained.

The outer side of the cylinder was free, that is, having no external load and displacement constraints, and the boundary conditions can be expressed as:

$$\sigma_r(1, \psi) = 0, \quad \tau_{rz}(1, \psi) = 0 \tag{2}$$

The Love displacement function ζ of the cylindrical axisymmetric problem was as follows (8):

$$\zeta = R^3 \left\{ \begin{aligned} & A_0 \frac{\kappa^3 \psi^3}{6} + C_0 \frac{\kappa \psi}{2} \rho^2 \\ & + \sum_{m=1}^{\infty} \frac{\sin(m\pi\psi)}{\eta_m^3} \left[\begin{aligned} & A_m \eta_m \rho \mathrm{I}_1(\eta_m \rho) \\ & + B_m \mathrm{I}_0(\eta_m \rho) \end{aligned} \right] \\ & + \sum_{s=1}^{\infty} \frac{\mathrm{J}_0(\xi_s \rho)}{\xi_s^3} \left[\begin{aligned} & C_s \sinh(\lambda_s \psi) \\ & + D_s \lambda_s z \cosh(\lambda_s \psi) \end{aligned} \right] \end{aligned} \right\} \tag{3}$$

where I_n was the n-order virtual Bessel function; J_n was the n-order Bessel function; $\eta_\mathrm{m} = \frac{m\pi}{\kappa}$, $\lambda_s = \xi_s \kappa$; ξ_s was the zero point of the first-order Bessel function.

The expression of the radial stress σ_r, the tangential stress σ_θ, the vertical stress σ_z, the radial displacement u, and the vertical displacement w of the cylinder (9) was as follows:

$$\left. \begin{aligned} & u = -\frac{1}{2G} \frac{\partial^2 \zeta}{\partial r \partial z}, \quad w = \frac{1}{2G} \left[2(1-v)\nabla^2 - \frac{\partial^2}{\partial z^2} \right] \zeta \\ & \sigma_r = \frac{\partial}{\partial z} \left(v\nabla^2 - \frac{\partial^2}{\partial r^2} \right) \zeta, \quad \sigma_\theta = \frac{\partial}{\partial z} \left(v\nabla^2 - \frac{\partial}{r \partial r} \right) \zeta \\ & \sigma_z = \frac{\partial}{\partial z} \left((2-v)\nabla^2 - \frac{\partial^2}{\partial z^2} \right) \zeta \\ & \tau_{rz} = \frac{\partial}{\partial r} \left((1-v)\nabla^2 - \frac{\partial^2}{\partial z^2} \right) \zeta \end{aligned} \right\} \tag{4}$$

where G and v were the shear modulus and Poisson's ratio, respectively, and $\nabla^2 = \left(\frac{\mathrm{d}^2}{\mathrm{d}r^2} + \frac{\mathrm{d}}{r\mathrm{d}r} \right)$ was the Laplacian operator.

Equation 3 can satisfy the symmetric boundary condition at $\psi = 0$, and the boundary condition at $\psi = 1$ was expanded as the Fourier-Bessel series of $\mathrm{J}_0(\xi_s \rho)$. The boundary condition at $\rho = 1$ was developed as the $\sin(m\pi\psi)$ Fourier series. Substituting boundary conditions (1), the following series of equations can be obtained:

$$\left.\begin{array}{l}
(1-2v)\kappa A_0 + 4\kappa(1-v)C_0 = -\tilde{w}_0 \\
vA_0 + (2v-1)C_0 + \frac{1}{2}\sum_{s=1}^{\infty}\left[(C_s + (2v+1)D_s)\Gamma_{s0} + D_s\Lambda_{s0}\right] = 0 \\
\left.\begin{array}{l}
\begin{bmatrix} A_m[(2v-1)\mathrm{I}_0(\eta_m) + \eta_m \mathrm{I}_1(\eta_m)] \\ -B_m\left[\mathrm{I}_0(\eta_m) - \frac{\mathrm{I}_1(\eta_m)}{\eta_m}\right] \\ + \sum_{s=1}^{\infty}\left[(C_s + (2v+1)D_s)\Gamma_{sm} + D_s\Lambda_{sm}\right] \end{bmatrix} = 0 \\
[2(1-v)\mathrm{I}_1(\eta_m) + \eta_m \mathrm{I}_0(\eta_m)]A_m + \mathrm{I}_1(\eta_m)B_m = 0
\end{array}\right\} \quad m = 1, 2, \cdots \\
\left.\begin{array}{l}
C_s\sinh(\lambda_s) + D_s\begin{bmatrix}\lambda_s\cosh(\lambda_s) \\ -2(1-2v)\sinh(\lambda_s)\end{bmatrix} = 0 \\
C_s\begin{bmatrix}k_\tau\xi_s\sinh(\lambda_s) \\ +\vartheta\cosh(\lambda_s)\end{bmatrix} + D_s\begin{bmatrix}(k_\tau\lambda_s\xi_s + \vartheta)\cosh(\lambda_s) + \\ (k_\tau\xi_s 2v + \vartheta\lambda_s)\sinh(\lambda_s)\end{bmatrix} \\
-\vartheta\sum_{m=1}^{\infty}\frac{(-1)^m\xi_s}{\eta_m}\begin{bmatrix}A_m U_{sm} + \\ B_m T_{sm}\end{bmatrix} - \vartheta\xi_s C_0\tilde{\rho} = 0
\end{array}\right\} \quad s = 1, 2, \cdots
\end{array}\right\} \quad (5)$$

where

$$\Gamma_{sm} = \frac{2\lambda_s(-1)^m\sinh(\lambda_s)}{\lambda_s^2 + (m\pi)^2};\ \Lambda_{sm} = \frac{2\lambda_s(-1)^m}{\lambda_s^2 + (m\pi)^2}\begin{bmatrix}\lambda_s\cosh(\lambda_s) \\ + \frac{(m\pi)^2 - \lambda_s^2}{\lambda_s^2 + (m\pi)^2}\sinh(\lambda_s)\end{bmatrix};$$

$$T_{sm} = \frac{-2\xi_s\mathrm{I}_1(\eta_m)}{(\eta_m^2 + \xi_s^2)\mathrm{J}_0(\xi_s)};\ U_{sm} = \frac{2\eta_m}{(\eta_m^2 + \xi_s^2)\mathrm{J}_0(\xi_s)}\left[\frac{2\eta_m\xi_s\mathrm{I}_1(\eta_m)}{\eta_m^2 + \xi_s^2} - \xi_s\mathrm{I}_0(\eta_m)\right];$$

$$\tilde{\rho} = -\frac{2}{\xi_s\mathrm{J}_0(\xi_s)};\ \tilde{w} = \frac{2Gw}{R};\ \vartheta = \frac{k_u R}{2G}.$$

The parameters ϑ in the above formulas are the non-dimensional coefficients that characterize the horizontal friction level of the end. If the radial Fourier-Bessel series takes m_0 terms, the vertical Fourier series takes s_0 terms, and then Equation 3 has $2(m_0 + s_0 + 1)$ constants: $A_0\ A_1\ \ldots\ A_{m0},\ B_1\ \ldots\ B_{m0},\ C_0\ C_1\ \ldots\ C_{s0},\ D_1\ \ldots\ D_{s0}$, while Equation 5 happens to have $2(m_0 + s_0 + 1)$ linear equations. Therefore, the linear equations of Equation 5 can be used to solve all the above-mentioned contingencies.

Strain and Strain Distribution of Cylinders

The specimens were subjected to uniform compression in varius conditins, namely: with no friction at the end of the cylinder, with vertical compressive stress $\sigma_0 = P/\pi R^2$, shear stress $\tau_0 = P/2\pi R^2$, and radial tensile strain $\varepsilon_0 = vP/E\pi R^2$. Under the same external load, the ratios of vertical compressive stress σ_z to σ_0, shear stress τ to τ_0, and radial tensile strain ε_r to ε_0 were denoted as λ_σ (vertical compressive stress ratio), λ_τ (shear stress ratio), and λ_ε (radial strain ratio), respectively. Figure 2 shows the distribution of the vertical compressive stress ratio λ_σ, the shear stress ratio λ_σ, and the radial strain ratio λ_ε of the cylinder with the Poisson's ratio $v = 0.25$, and the height-to-diameter ratio $\kappa = 1$ (Figure 2).

As can be seen from Figure 2, when the end surface of the cylinder was constrained, the compressive stress of the cylinder was inhomogeneous. The compressive stress on the central axis increased from the top to the middle, and the stress concentration occurred near the edge of the end surface; $\lambda_\sigma \approx 1.042$ at the center of the symmetry plane, $\lambda_\sigma \approx 0.96$ at the edge, and the difference was about 8.5%. The shear stress increased gradually from the end to the center on the central axis, gradually increased from the middle to the side, and the maximum appeared at the middle edge, but $\lambda_\tau < 1$, that is, the end face constraint will reduce the shear stress. Because of the end restraint effect, the λ_ε of the cylinder showed a concave curve, the maximum tensile strain was at the middle edge of the specimen, and was greater than the tensile strain compared with the uniform compression state, $\lambda_\varepsilon \approx 1.17$.

Figure 3 shows the maximum radial tensile strain ratio $\boldsymbol{\lambda}_{\varepsilon\max}$ (appearing at the midpoint of the side) of the cylinder when the radial displacement of the end face was completely constrained with the condition of the different height-to-diameter ratio κ and Poisson's ratio v (Figure 3).

It can be seen from Figure 3 that the maximum radial tensile strain ratio $\boldsymbol{\lambda}_{\varepsilon\max}$ of the cylinder increased with the increase of v and decreased with the increase of κ. $\boldsymbol{\lambda}_{\varepsilon\max}$ increased from 1.21 to 1.01 when κ increased from 0.5 to 2 and $v = 0.4$; $\boldsymbol{\lambda}_{\varepsilon\max}$ increased from 1.05 to 1.13 when v increased from 0.1 to 0.4, and $\kappa = 1$.

The influence of the different confinement levels on the maximum radial tensile strain $\varepsilon_{r.\max}$ of the cylinder was characterized by the introduction of a tensile strain friction coefficient χ_ε, which was defined as:

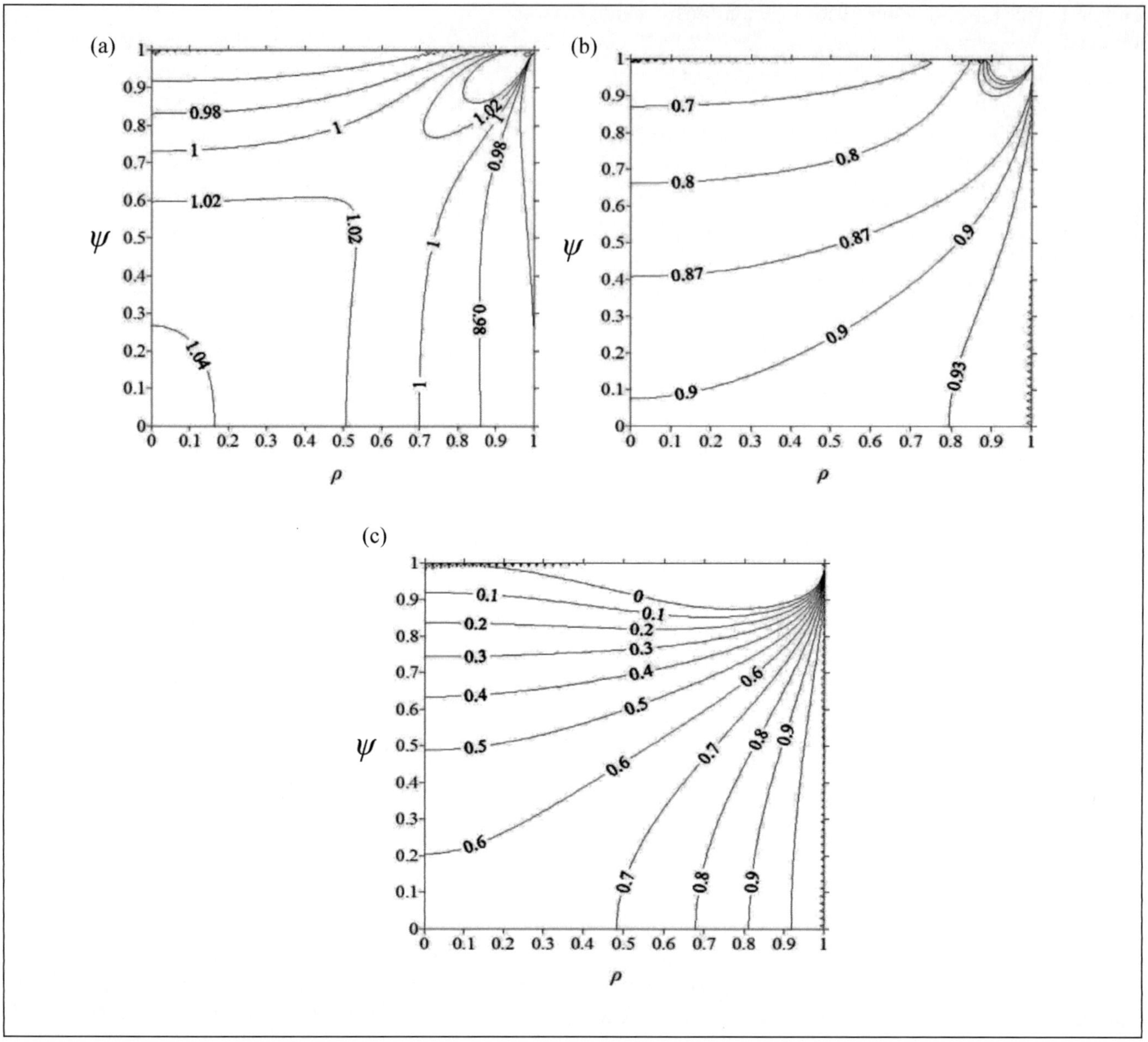

Figure 2. Contour map of cylinder stress (strain): (*a*) λσ, (*b*) λτ, (*c*) λε.

$$\chi_{\varepsilon} = \frac{\varepsilon_{r.\max} - \varepsilon_0}{\varepsilon_{\infty.m} - \varepsilon_0} \tag{6}$$

where $\varepsilon_{\infty.m}$ was the maximum radial tensile strain of the cylinder when the radial displacement of the end face was completely confined.

The friction coefficient χ_{ε} was related to the Poisson's ratio v and the height to diameter ratio of the cylinder (κ), and the horizontal friction coefficient of the end face (ϑ). When the height to diameter ratio of the cylinder is greater, namely a more slender cylinder, the influence of end constraint is weak. The relationship between χ_{ε} and v, ϑ is shown in Figure 4, when the height-to-diameter ratio $\kappa = 1$, which was the most common condition (Figure 4).

As can be seen from Figure 4, the smaller the v, the more the curve was upward, meaning that the stress–strain state of the cylinder was closer to the state when the end was completely confined. When the material Poisson's ratio v and the end horizontal friction coefficient ϑ are known, the maximum tensile strain $\boldsymbol{\lambda}_{\varepsilon.\max}$ of the cylinder can be obtained by using Figures 3 and 4.

Vertical Deformation of the Cylinder

Due to the end restraint, the amount of external load P required by the cylinder to produce the same amount of w_0 is increased. The ratio of the external load P required

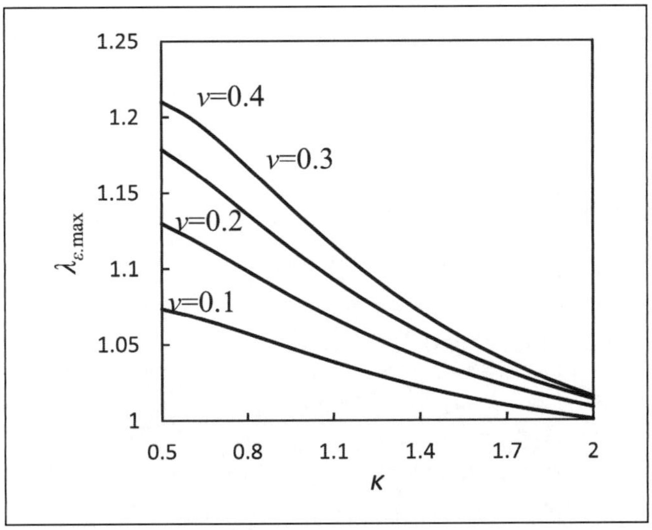

Figure 3. The relationship between λ_ε and κ,v.

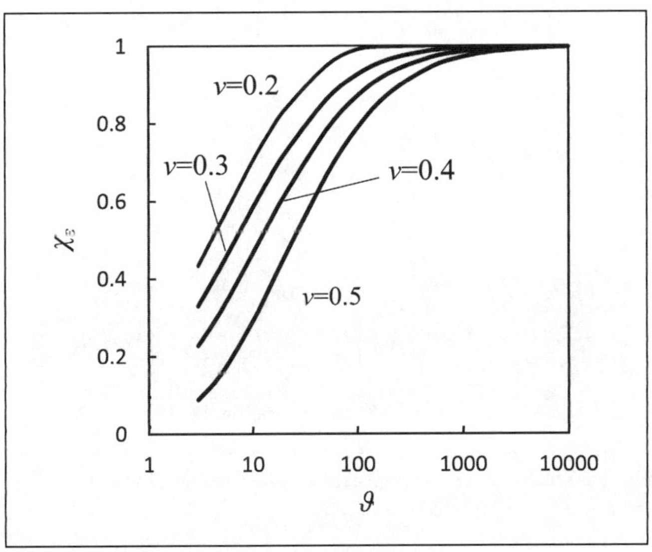

Figure 4. The relationship between χ_ε and ϑ, v.

to produce the unit displacement at the end of the cylinder to the external load P_0 required in the case of uniform pressure was denoted by λ_P, and λ_P was the load ratio. The load ratio λ_P increases rapidly with the increase of the Poisson's ratio v, and the greater the horizontal friction coefficient at the end, the faster the growth. When the height to diameter ratio of the cylinder is 1 ($\kappa = 1$), and the end is completely constrained, $v = 0.25$, $\lambda_P = 1.08$; $v = 0.4$, $\lambda_P = 1.72$, as shown in Figure 5*a*. The smaller the height-to-diameter ratio, the more significant the influence of the end restraint on the load ratio. When $v = 0.4$, and the end is completely constrained, $\lambda_P = 1.54$ when $\kappa = 0.5$; $\lambda_P = 1.15$ when $\kappa = 2$, the difference is about 33.9%, see Figure 5*b*.

It can be deduced that the compressive elastic modulus of the material (the relative compressive stress of the cross section due to the external load × the height of the column/the relative displacement between the two ends of the cylinder) is measured by the top surface method if the local compression caused by the uneven surface is excluded, and the partial amount of the value of elastic modulus will be equal to the load ratio λ_P. In fact, the effect of local compression caused by the unevenness of the end face tends to exceed the latter, so that the compressive modulus of the measured material measured by the top method is mostly small (Figure 5).

In order to overcome the shortcomings of the top surface method, the newly established Asphalt Pavement Design Code for P.R.China (*10*) recommends measurement of the compressive elastic modulus of the material by the lateral surface method, namely, the compressive modulus of the material is calculated by measuring the relative displacement of the outer surface of the cylinder (the average compressive stress of the cross section caused by the external load × the displacement between the two measuring points/the relative displacement between the two measuring points). The lateral surface method can effectively eliminate the adverse effects of local compression caused by the uneven surface, but the effect of the horizontal friction of the end face was still very obvious.

Figure 6 shows the relationship between the ratio of elastic modulus E to the true elastic modulus E_0 and the Poisson's ratio v determined by the lateral method with three kinds of height-to-diameter ratio of $\kappa = 1.0$, 1.5, and 2, as recommended by the specification, when the end was fully restrained (Figure 6).

As can be seen from Figure 6, the E/E_0 was >1, namely, when there was friction at the end, the elastic modulus E measured by the lateral surface method was rather large, and its deviation increased with the increase of the end friction coefficient ϑ and the Poisson ratio v, and decreased with the increase of the height to diameter ratio κ. For cement concrete with Poisson's ratio $v \leq 0.2$, the partial amount of the compressive elastic modulus can be controlled within 3%, but for asphalt concrete with Poisson's ratio $v \geq 0.4$, the partial amount of the elastic modulus may be up to 1/4 to 1/3. It was difficult to meet the engineering requirements, or the friction at the end surface of the specimen should be reduced, or a new elastic modulus measurement method was needed.

Failure Mechanism of Unconfined Compression Specimen

Several studies show that the dimensions of the end constraints can affect the failure morphology of the unconfirmed compression specimen (*4, 5, 7*). If the end

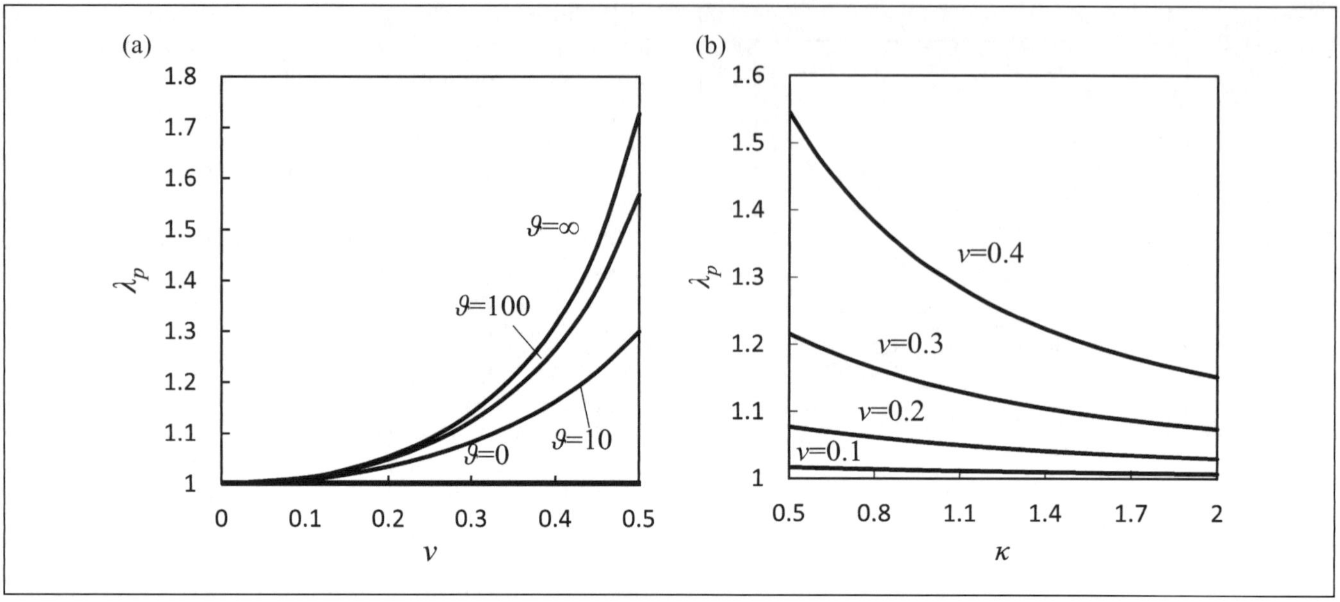

Figure 5. The load ratio under different conditions: (*a*) $\kappa=1$, (*b*) $\vartheta=\infty$.

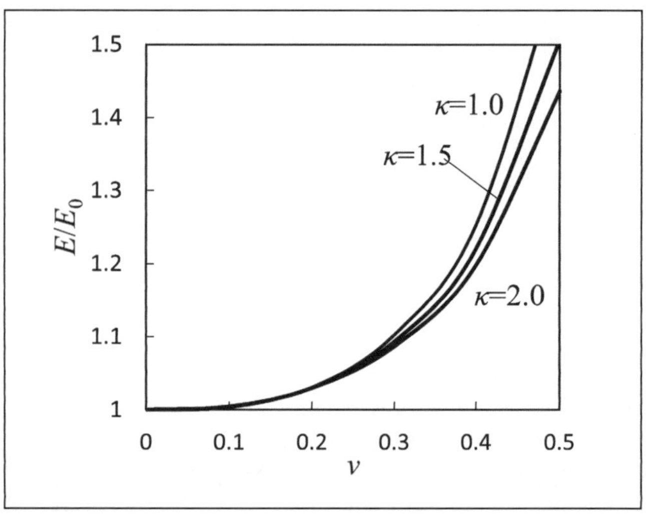

Figure 6. The compressive modulus deviation.

Figure 7. Unconfined compression failure mode of specimen: (*a*) reduced friction, (*b*) no treatment.

constraint was large, the specimen was broken after the double cone shape; and if the face constraint was small, the specimen formed a number of top-down cracks. Indoor was used to verify this conclusion. The weak end constraints were achieved by placing two pieces of polyfluortetraethylene plate and lubricating oil between them. The friction coefficient was very small (0.02) with this treatment. The specimen was a cement stabilized soil with a 10% cement content. The height and diameter were 50 mm. After seven days, the unconfined compressive strength test was carried out. The failure morphology of the specimen is shown in Figure 7. It was found that with the increase of the load, the specimen gradually expanded outside, and small vertical cracks were formed in the vicinity of the end of the specimen, and then the crack extended up and down to break through. When the end surface was not treated to reduce friction, with increased load, the middle part of the specimen gradually bulged and small network cracks formed, followed by crack development and expansion, and finally bearing capacity was lost. These test results were the same as those of previous studies (*4*) (Figure 7).

From the initiation and propagation of cracks and failure modes of the specimen, the failure of the unconfined compression specimen conforms to the second strength theory, that is, the material tensile strain caused by external loads exceeds the ultimate tensile strain. If the end face is weakly restrained, the tensile strain of the specimen is almost the same everywhere, and the crack is caused from both ends because of the structural defects

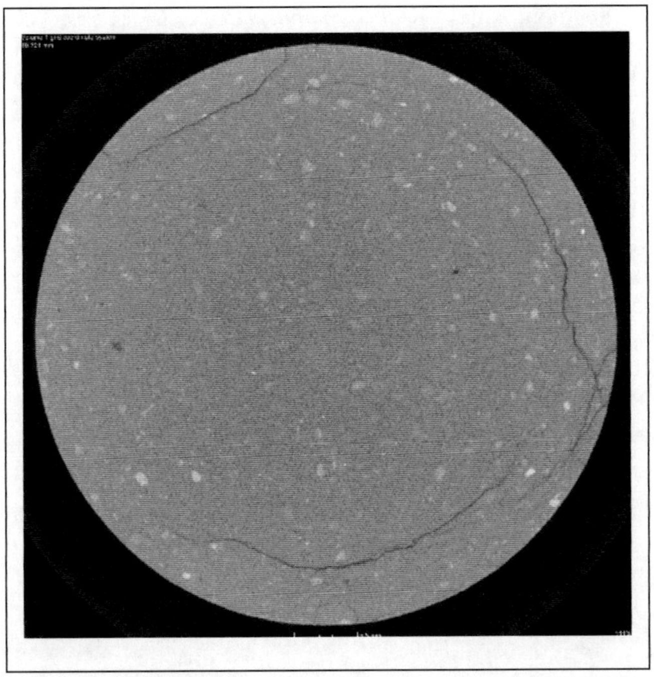

Figure 8. Cross-sectional scanning image of cylinder quartile.

at the end. If the end face constraint is large, the maximum tensile strain of the specimen occurs in the middle side, and the crack is first generated. At the beginning of the crack initiation, the specimen is close to its ultimate bearing capacity. Figure 8 shows a CT scan of the section of the specimen at the specimen height of one-quarter. As can be seen from Figure 8, the crack expanded inward from the outside, and as the cracks developed and penetrated rapidly, the bearing capacity of the specimen was gradually lost. If the end face is constrained, the specimen is more susceptible to damage due to $\varepsilon_{r.\max} > \varepsilon_0$, that is to say, the compressive strength of the material to be measured is too weak (Figure 8).

Conclusions

For the cylindrical unconfined compression test, the elastic axisymmetric mechanical model of the cylinder with horizontal friction was constructed, and the elastic solution was deduced. The cylindrical body with different cylinder sizes and end-face friction was calculated. The experimental results of the unconfined compression of the cylinder were studied by laboratory testing. The main conclusions are as follows:

1. When the radial displacement of the cylindrical end is constrained, the stress and the strain of the cylinder are no longer uniform, and the inhomogeneity increases with the decrease of the height to diameter ratio of the cylinder and the increase of the Poisson's ratio v.
2. When the end is restrained, the maximum tensile strain appears in the middle of the side of the specimen, and its value increases with the restraint effect.
3. If the end of a cylinder was constrained, the vertical displacement of the same needed more load, so the compressive modulus according to the measured method was too large, the partial amount of compressive elastic modulus of cement concrete can be controlled within 3%, the elastic modulus of asphalt concrete measured values may be too large between 1/4 and 1/3.
4. From the initiation and propagation of cracks and failure modes of the specimen, the failure of the unconfined compression specimen conforms to the second strength theory. When the end constraint was large, it first formed a small network crack in the middle of the outer side of the cylinder, which was double-tapered when damaged. If the end face was weakly restrained, the cracks penetrated up and down.

Acknowledgments

This research was financially supported by the National Natural Science Foundation of China (Grant No. 51778479), the National Key R&D Program of China (Grant No. 2016YFB0303200), and the International Exchange Program for Graduate Students (Tongji University).

Author Contributions

The authors confirm contribution to the paper as follows: study conception and design: Jian Xiao, Zhiming Tan, Jingliang Chen; data collection: Jian Xiao; analysis and interpretation of results: Jian Xiao, Zhiming Tan; draft manuscript preparation: Jian Xiao, Zhiming Tan, Jingliang Chen. All authors reviewed the results and approved the final version of the manuscript.

References

1. National Standard of the People's Republic of China. *Standard for Test Method of Mechanical Properties on Ordinary Concrete*. GB/T 50081-2002. China Architecture and Building Press, Beijing, 2003.
2. Profession Standard of the People's Republic of China. *Standard Test Methods of Bitumen and Bituminous Mixtures for Highway Engineering*. JTG E20-2011. People's Communication Press, Beijing, 2011.
3. Profession Standard of the People's Republic of China. *Test Methods of Soils for Highway Engineering*. JTG E40-2007. People's Communication Press, Beijing, 2007.
4. Filon, L. N. G. On the Elastic Equilibrium of Circular Cylinders Under Certain Practical Systems of Load. *Philosophical Transactions of the Royal Society A: Mathematical*

Physical & Engineering Sciences, Vol. 198, 1902, pp. 147–233.
5. Beisetaev, R. B, and V. S. Nikiforovskii. Strengths of Solids Under Uniaxial Compression. *Soviet Mining*, Vol. 12, 1976, pp. 244–248.
6. Saito, H. The Axially Symmetrical Deformation of a Short Circular Cylinder. *Transactions of the Japan Society of Mechanical Engineers*, Vol. 18, 1952, pp. 21–28.
7. Watanabe, S. Elastic Analysis of a Axi-Symmetric Finite Cylinder Restrained to Radial Displacement in Loading Ends. *Proceedings of the Japan Society of Civil Engineers*, Vol. 20, 1992, pp. 85–94.
8. Chau, K. T., and X. X. Wei. Finite Solid Circular Cylinders Subjected to Arbitrary Surface Load. Part I: Analytic Solution. *International Journal of Solids & Structures*, Vol. 37, 2000, pp. 5707–5732.
9. Xu, Z. Elasticity, 4th edition, Higher Education Press, Inc., Beijing, 2006.
10. Profession Standard of the People's Republic of China. *Specifications for Design of Highway Asphalt Pavement*. JTG D50-2016. People's Traffic Press, Beijing, 2016.

The Standing Committee on Basic Research and Emerging Technologies Related to Concrete (AFN10) peer-reviewed this paper (18-02592).

The opinions expressed in this paper are those of the authors and not necessarily those of the sponsors.

Article

TRR
JOURNAL OF THE TRANSPORTATION RESEARCH BOARD

Transportation Research Record
2018, Vol. 2672(27) 45–55

DOI: 10.1177/0361198118787983
journals.sagepub.com/home/trr
SAGE

Use of Fiber-Reinforced Self-Consolidating Concrete to Enhance Serviceability Performance of Damaged Beams

Haider A. Abdulhameed[1,2], Hani Nassif[1], and Kamal H. Khayat[3]

Abstract

The use of fiber-reinforced self-consolidating concrete (FR-SCC) in repairing damaged concrete beams has been evaluated. An experimental program was conducted to design and test key fresh and hardened properties of SCC and FR-SCC mixtures. The designed FR-SCC mixtures included two types of supplementary cementitious materials (silica fume (SF) and slag (SL)) and two types of fibers (steel fiber (STF) and polypropylene fiber (PPF)) were used. To ensure good workability to repair congested areas, the optimized volume fractions of the STF were 0.25% and 0.50% compared with 0.10%, 0.15%, and 0.20% for the PPF. In addition, the flexural behavior of 10 beam specimens was investigated. The main reinforcement for the control beams consisted of #5 reinforcing bars, while the main reinforcement for the repaired beams was either #4 or #3 reinforcing bars that were introduced to simulate 35% and 65% reduction of the bar areas, respectively, due to corrosion. The results demonstrate that the optimized FR-SCC mixtures are effective repair materials and can develop adequate bond strength to existing concrete. The flexural test results showed that the repair mixtures were able to increase the cracking load for the repaired beams compared with the control beams. Such an increase is expected to contribute to extending the life of the damaged member or structure at the service load level. This paper also presents a comparison of the predicted values for the first-crack load strength using the ACI 544 code equation with the experimental data. Results showed that the code equation provides safe prediction.

Poor durability design and combined with a lack of periodic maintenance and repair can result in expensive maintenance and even failure of the structure causing potential human catastrophes. Repairs should increase the lifetime of a structure by restoring structural capacity and extending the service life of the structure. Repairs are used to avoid or postpone demolition and reconstruction of a structure for economic purposes. A report in 2017 by the American Society of Civil Engineers (ASCE) estimated that $4.59 trillion would be needed by 2025 to upgrade the condition of infrastructure in the United States from Grade D (Poor) to Grade B (Good) (*1*). Therefore, repairs need to be of high quality, exhibit enhanced durability, and they need to be applied efficiently to avoid repeat cases of multiple repairs.

The efficiency of a given repair depends on both the repair material quality and the ability of the material to fill the section being repaired, especially in congested areas. Repair efficiency is also a function of repair method, compatibility, and the bond between the repair material and the substrate. To satisfy these requirements, repair materials should be carefully selected to ensure a successful repair. The function and performance of buildings, bridges, or pavements can be improved by ensuring good repairs. Unfortunately, poor repairs can fail prematurely or can even accelerate the deterioration of the repaired concrete structure. Poston et al. (*2*) have shown that the choice of appropriate repair materials is crucial to material properties and the behavior of the composite repaired section under service exposure conditions. Previous studies (*3–5*) have proven that the evaluation of the repair materials cannot be done only on the mechanical properties. Bond strength and compatibility tests are also important tests to prove that a repair has been successful.

Self-consolidating concrete (SCC) has been successfully used for the rehabilitation and repair of reinforced concrete structures (*6*). SCC mixtures usually have a higher content of

[1]Rutgers Infrastructure Monitoring and Evaluation (RIME) Group, The State University of New Jersey, Department of Civil and Environmental Engineering, Piscataway, NJ
[2]Building and Construction Engineering Department, University of Technology, Baghdad, Iraq
[3]RE-CAST Tier I University Transportation Center (UTC), Center for Infrastructure Engineering Studies, Rolla, MO

Corresponding Author:
Address correspondence to Haider A. Abdulhameed: haider.adel@rutgers.edu

fine materials and paste than normal concrete, which can increase shrinkage and the risk of cracking (7). It is well known that repairs done using concrete are especially susceptible to cracking as a result of shrinkage, including restrained shrinkage. To combat this frequent occurrence, fiber reinforcement has been increasingly used to improve tensile and flexural strength while reducing cracking density and crack opening. Fiber-reinforced SCC (FR-SCC) can be placed easily compared with conventional fiber-reinforced concrete of normal consistency.

The main objective of this paper is to investigate the mechanical properties of FR-SCC intended for the repair of flexural concrete members. The bond strength and compatibility of FR-SCC as a repair material is investigated. The structural behavior of full-scale repaired beams with simulated bar sections reflection corrosion reduction of the steel and concrete cover lost is evaluated. The goal of using FR-SCC is to improve the cracking capacity of the damaged beams so that the service load can be carried safely. The ACI 544 (8) equation for predicting the first-crack strength for beams repaired with FR-SCC was also studied.

Experimental Procedures

Concrete Mixtures and Fresh Properties

Class A concrete specified by the New Jersey Turnpike Authority (NJTA) with a minimum compressive strength of 4,500 psi at 28 days was selected for the substrate concrete. The ordinary Portland cement conforming with ASTM C150 Type I cement was used for all types of concrete. The nominal maximum size of the coarse aggregate is 3/4 in. for Class A concrete and 3/8 in. for the SCC and FR-SCC. Both coarse aggregates were crushed granite, conforming to ASTM C33. A densified silica fume with a specific gravity of 2.10–2.40 was used. A Grade 120 slag with a specific gravity of 2.90 was employed. The Blaine fineness of the cement and slag (SL) are 380 and 594 m^2/kg, respectively, and the specific surface area of the silica fume (SF) determined from BET analysis is 22,280 m^2/kg. The steel fiber (STF) used is macro crimped fiber with a length of 1.5 in. The polypropylene fiber (PPF) is micro 1/4 in. in length. The mixture proportions of the SCC were based on a mixture design carried out by Brown et al. (9) with some modifications. The details of identification and mixture proportions of the 10 investigated mixtures are listed in Table 1.

The fresh properties of the optimized mixtures are also presented in Table 1. The use of PPF exhibited problems in passing ability when incorporated at higher volumes of 0.2% and 0.25%, and especially in mixtures using SF. The minimum slump flow value was set at 21 in., whereas the maximum obtained was 25 in. During the slump flow test, the T20 was measured. The results ranged between 4 and 8 sec. It is well known that the viscosity increases with the increase of fiber content. For repair purposes, a low viscosity is desired with a T20 value around 5 sec. Most of the mixtures exhibited high stability with no sign of bleeding or segregation (VSI = 0). For the 35SL25S and 35SL50S mixtures, a halo of mortar was seen on the outer edge of the concrete slump flow (VSI = 1). According to ASTM C1611M–14, a value for VSI equal to 1 is considered stable SCC or FR-SCC.

Modified L-box and J-ring tests were used to evaluate the passing ability of the SCC or FR-SCC. The typical L-box is designed for the SCC, therefore, one rebar instead of three was used in the modified L-box for the FR-SCC (10). The test results for the L-box showed looking ratio values greater than 0.8 which is an excellent passing ability. The difference between the slump flow and the J-ring readings was also taken as an indicator of passing ability. For SCC, a difference of 1 in. is considered as no visible blocking and adequate passing ability. The difference in flow has increased with the increase in fiber content and it is not, therefore, recommended to use the mixes with high fiber volume content in repairing the very congested reinforcement areas unless a partial compaction is to be applied. Khayat et al. (10) recommended using a modified J-ring of 8 bars instead of 16 for FR-SCC.

Third-Point Loading Composite Prism Test (Compatibility)

A test method to examine the compatibility of the repair material with the substrate was developed by Czarnecki et al. (11). The concrete prisms measure 16 in. in length with a cross-section of 4 x 4 in. and were cast in accordance with ASTM C 78. Figure 1 shows the sample dimensions, including the repair part and the test setup. After demolding the substrate specimens, the prisms were moist cured for 28 days, and then the wide-mouthed notch areas were textured using a dry brushing. The rough surface textured substrate specimens were air-dried for 7 days before batching the notched area with the repair FR-SCC materials. The composite samples were demolded after 24 hours and cured until the time of testing.

Bond Strength Test

The bond strength of the repair materials was determined in accordance with ASTM C882. The repair material was bonded to a substrate concrete specimen on a slanted elliptical plane inclined at 30° angle from vertical to form a 3 × 6 in. composite cylinder, as shown in Figure 2. Before bonding the repair material to the substrate concrete, the slanted surface of the substrate concrete specimen was cleaned and dried. The curing procedure used was according to ASTM C882 and as explained in the third point loading composite prism test. The test is performed by determining the compressive load required to fail the composite cylinder, and the bond strength was calculated by the maximum applied load divided by the area of slant surface. According to Pattnaik

Table 1. Mixture Proportions for One Cubic Yard and Fresh Properties of SCC and FR-SCC Mixes

		Composition										Fresh properties						
Mixture	Identification	Cement, lb	SL or SF, lb	Total binder weight, lb	Water, lb	w/cm	Coarse aggregate (#8), lb	Fine aggregate (sand), lb	AEA, oz./cwt	HRWRA, oz/cwt	MRWRA, oz/cwt	Slump flow, in.	T20, S	VSI	J-Ring, in.	Modified L-Box, h2/h1	Air content, %	Difference in flows, in.*
Class A	Class A	611	0	611	259	0.420	1800	1275	4.9	0.0	24.4	–	–	–	–	–	–	–
35SL	35% SL	439	236	675	287	0.425	1436	1436	0.5	7.0	0	23.5	4.0	0	22.5	1.00	7.0	1.0
35SL25S	35% SL + 0.25%STF									7.5		21.5	5.5	1	20.0	1.00	6.5	1.5
35SL50S	35% SL + 0.50%STF									8.5		22.0	6.5	1	19.0	0.90	6.5	3.0
35SL15P	35% SL + 0.15%PPF									10.0		23.5	4.5	0	19.0	0.90	6.5	3.5
35SL20P	35% SL + 0.20%PPF									11.0		23.0	4.5	0	18.0	0.80	6.0	5.0
10SF	10% SF	603	68							8.5		25.0	5.0	0	21.5	1.00	6.5	3.5
10SF25S	10% SF + 0.25%STF									12.3		25.0	5.3	0	21.0	0.95	6.5	4.0
10SF50S	10% SF + 0.50%STF									22.0		24.0	6.0	0	19.0	0.93	6.0	5.0
10SF10P	10% SF + 0.10%PPF									17.2		23.0	5.0	0	18.0	0.90	6.5	5.0
10SF15P	10% SF + 0.15%PPF									26.5		21.0	8.0	0	17.0	0.80	6.0	4.0

*Difference in flow = slump flow – J-Ring.

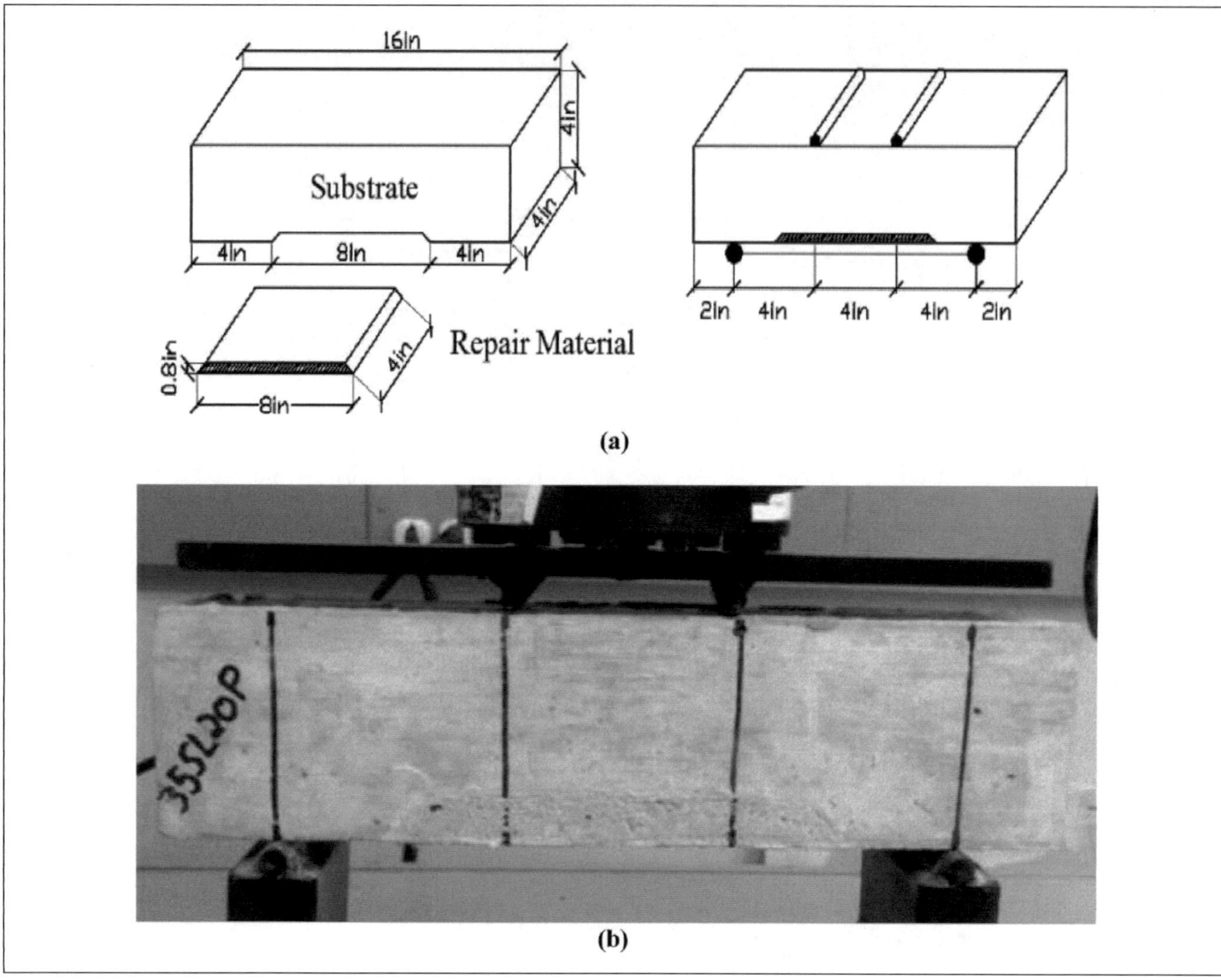

Figure 1. Compatibility test: (*a*) dimensions; (*b*) test setup.

(*12*), there are two classes of bond strength depending on the failure mode. If the mode of failure is an interface, an actual bond strength can be determined. Otherwise, a minimum bond strength can be calculated according to ASTM C882.

Preparing Full-Sale Beams for Flexural Strength Test

A total of 10 beams were used, eight of which were partially cast and repaired with FR-SCC optimized mixtures, while the other two were cast monolithically as control beams. This test can simulate repaired structural bridge elements. The main tensile reinforcement for the control beams was #5 bars, while the main reinforcement for the repaired beams were #4 and #3 beams. The smaller cross section rebars was employed to simulate the reduction of the reinforcing steel due to corrosion to the extent of 35% and 65% for the #4 and #3 rebars, respectively. For the top and shear reinforcement, two #2 and #3 rebars were used.

The two control beams and the substrate of the eight other beams were cast using a Class A concrete. Immediately after casting and vibrating the substrate concrete, a surface retarder was applied to delay the setting of the surface of the substrate. Beams were covered with a plastic sheet till the next day, and then a water stream was applied to remove some of the mortar and expose the aggregate to ensure a good bond of the substrate with the repair material. The beams were cured using wet burlaps for seven days and then continued to be air cured inside the lab until the day of repair. The day before the repair, the rebars were cleaned of any rust and the substrate surfaces were then cleaned of any dust and wetted to secure a saturated surface dry surface. After applying the repair, samples were covered with plastic sheeting. After 24 hours, samples were carefully demolded, and wet burlap cured for one day and tested at the age of three days from the day of repair.

Figure 3 shows the beam design details, test setup, and instrument locations. Various types of sensors, such as foil strain gauge, vibrating wire strain gauge (VWSG), and linear

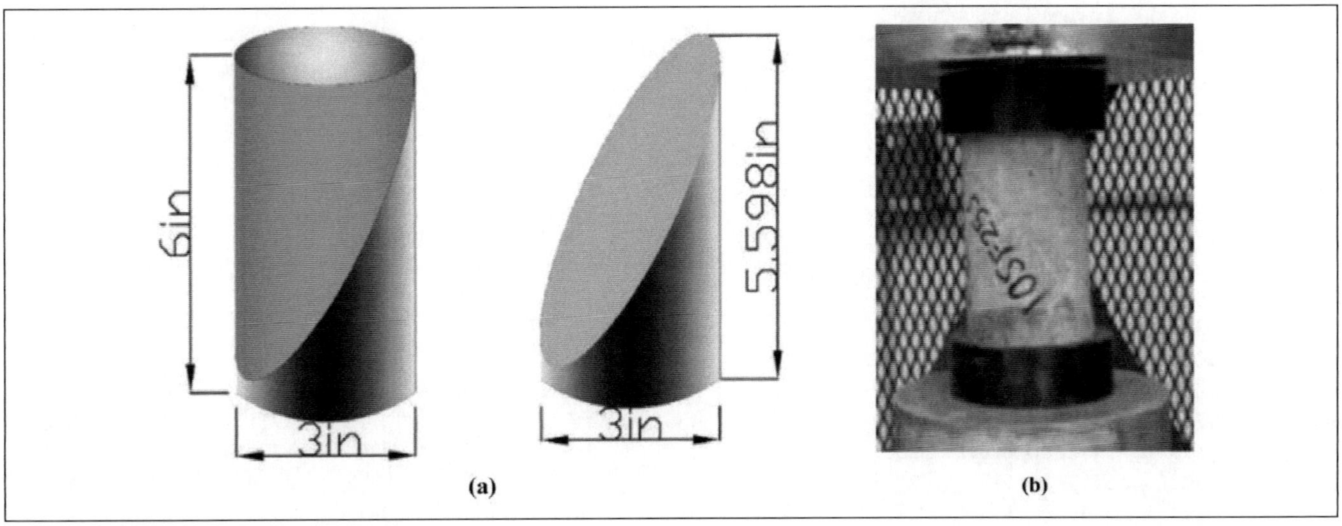

Figure 2. Slant shear bond strength test: (*a*) dimensions; (*b*) test setup for bond test.

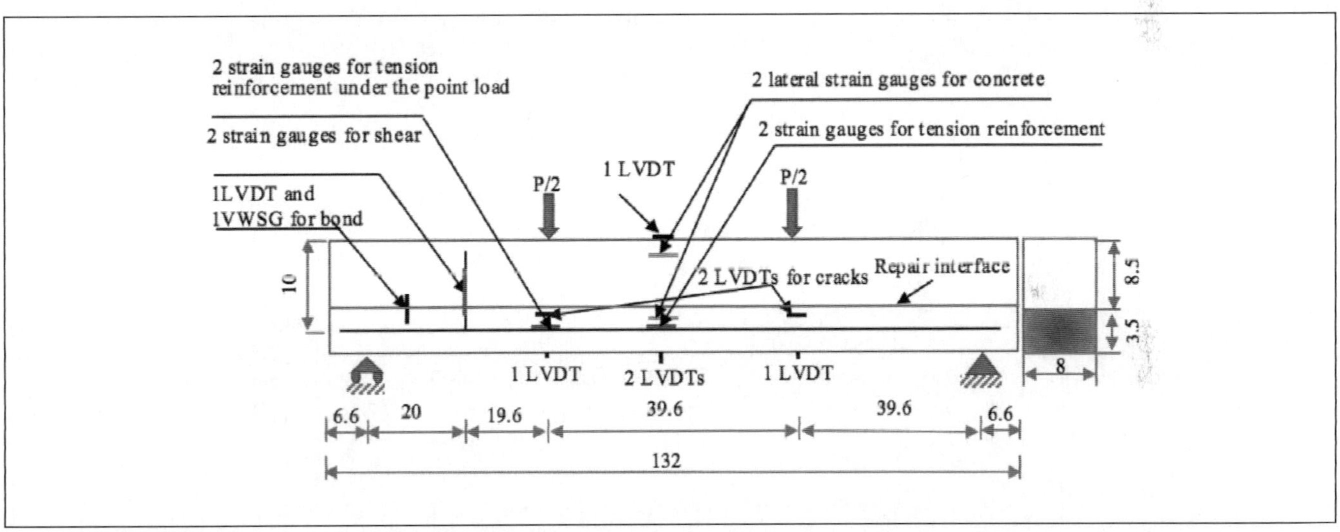

Figure 3. Schematic drawing of beam design details, test setup, and instrument locations.
*All dimensions are in inches.

variable differential transformer (LVDT), were instrumented in different locations to record the structural behavior using a simple beam with third-point loading. All data were collected and stored using one data acquisition system.

Test Results and Discussion

Hardened Properties

The summary of the hardened test results is summarized in Table 2. All samples were wet cured until 14 days. At 28 days, the compressive strength values for the SCC and FR-SCC were between 5,316 and 9,130 psi which is greater than the designed compressive strength of 5,000 psi. In general, all of the strength properties for the FR-SCC mixtures were enhanced compared with SCC mixtures. Figure 4 illustrates the free shrinkage test results. Mixtures made with 35% SL showed lower shrinkage than those with 10% SF. This is mainly attributed to the high fineness of the SF material. The results also demonstrated that the incorporation of both types of fibers reduced the shrinkage significantly compared with mixtures without fibers. It is worth mentioning that 35SL50S mixture showed the lowest shrinkage with 37% reduction compared with the 35SL mixture at 56 days. Emmons (*13*) pointed out that if the repair material shows a drying shrinkage in excess of 0.05% and 0.1% for moderate and high levels, respectively, an early failure could happen. The drying shrinkage results for the mixtures under study ranged between 0.013% and 0.032% at 28 days, which are below the moderate level.

Table 2. Hardened Properties Test Summary

	Compressive strength, psi (ASTM C39)			Splitting tensile strength, psi (ASTM C496)			Flexural strength, psi (ASTM C78)			Modulus of elasticity, ksi (ASTM C469)		
	Age-day			Age-day			Age-day			Age-day		
Mixture	1	3	28	1	3	28	1	3	28	1	3	28
Class A	3718	4498	5653	362	434	459	334	411	603	4121	4276	4591
35SL	2006	2972	5316	194	249	428	296	452	602	1491	3034	4000
35SL25S	2633	3848	5537	290	362	476	415	770	1036	1614	3584	4258
35SL50S	2309	3424	4976	368	422	568	563	924	1288	1928	3809	4427
35SL15P	2020	3410	5573	228	359	552	350	495	892	2674	3357	3850
35SL20P	1468	2426	4671	199	285	409	436	609	977	1683	2690	3534
10SF	2216	3750	5553	258	346	456	418	567	677	2751	2995	3920
10SF25S	3490	5414	8201	363	520	631	569	862	1076	3377	3912	5192
10SF50S	4300	6430	9130	476	610	777	865	1022	1346	3638	4558	5512
10SF10P	3722	4748	8134	370	493	587	461	631	837	3971	4310	5047
10SF15P	3676	4390	7935	324	427	652	544	703	919	3376	3894	4776

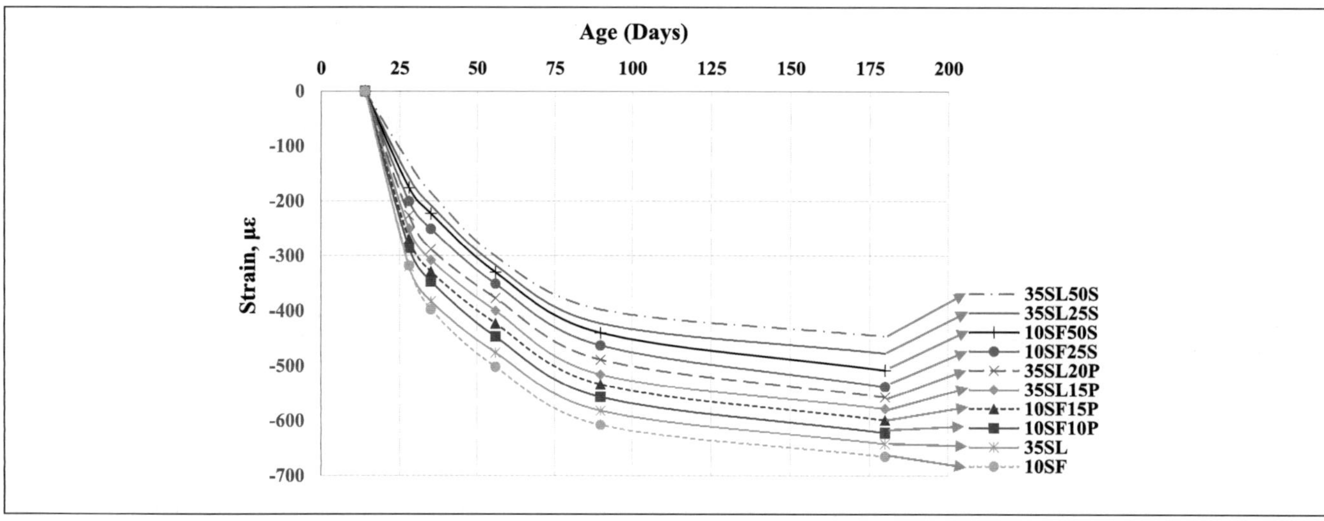

Figure 4. Free shrinkage at different ages according to ASTM C157.

Flexural Strength of Composite Samples (Compatibility)

Despite the fact that the flexural compatibility is not a standard test, it gives a good indication of how the small-scale composite repaired beam behaves under flexural loading by inducing tensile stresses in the repair material. The values of flexural strength for the substrate, repair mixtures, and compatibility that were tested separately are presented in Table 3. Results indicate that all composite prisms exhibited higher flexure strength compared with the substrate with a maximum increase of 177, 177, and 104% at 1, 3, and 28 days, respectively. In general, the failure location was in the middle, which reflects good compatibility. Depending on the point of failure determined through this test, a concrete sample's compatibility with a repair material is determined.

Czarnecki et al. (*11*) have found that if a failure crack appears through the repair material and the substrate, or through only the substrate, the two materials are compatible. If, however, a failure crack appears through the bonded area between the substrate and repair, the two materials are incompatible. Figure 5 shows the two types of failure mode observed. The good compatibility of the optimized repair mixtures during this study is mainly due to the low drying shrinkage of these materials.

Bond Strength

The results of the bond test in terms of the slant shear strength for the various types of concrete mixtures at 1, 3, and 28 days are presented in Table 4. Results demonstrate that the increase in fiber content increased the bond

Table 3. Results of Third Point Loading Composite Prism Test

Age-day	Repair material	Flexure strength, psi			Difference, %*	Failure location
		Control (Class A)	Repair material	Composite		
1	35SL25S	334	415	783	134	Center
	35SL50S		563	916	174	Center
	35SL15P		350	684	105	Center
	35SL20P		436	709	112	Center
	10SF25S		569	720	116	2 center and 1 edge
	10SF50S		865	925	177	Center
	10SF10P		461	753	125	Center
	10SF15P		544	759	127	Center
3	35SL25S	411	770	864	110	Center
	35SL50S		924	1021	148	Center
	35SL15P		495	781	90	Center
	35SL20P		609	890	117	Center
	10SF25S		862	781	90	2 center and 1 edge
	10SF50S		1022	1138	177	Center
	10SF10P		631	831	102	Center
	10SF15P		703	836	103	Center
28	35SL25S	603	1036	984	63	Center
	35SL50S		1288	1138	89	Center
	35SL15P		892	960	59	Center
	35SL20P		997	988	64	Center
	10SF25S		1076	1019	69	2 center and 1 edge
	10SF50S		1346	1232	104	Center
	10SF10P		837	1014	68	Center
	10SF15P		919	1028	70	Center

$$\text{*Difference} = \frac{\textit{Composite flexural strength} - \textit{substrate flexural streength}}{\textit{substrate flexural strength}} \times 100\%$$

strength. Pattnaik (*12*) reported that the strength ratios governed the mode of failure of the ready-mix repair materials. These ratios are compressive strength ratio and splitting tensile strength ratio tested separately (i.e., strength of repair material/strength of substrate). Table 4 also shows the strength ratios and the failure modes. It was observed that most of the mixtures failed with the substrate and repair mode with compressive strength ratio greater than or equal to 0.71. This behavior is mainly due to the high bond strength rather than the composite materials strength. The second noticed failure mode was the repair material failure when the compressive strength ratio is less than or equal to 0.62 and the splitting strength ratio is relatively high. The last observed failure mode was the interface and repair as in mixture 35SL20P with compressive and splitting tensile strength ratios of 0.39 and 0.61, respectively. In this case, the low compressive strength and splitting tensile for the repair material compared with the substrate led to such a failure. Figure 6 shows that the observed failure mode for most of the samples was in the substrate and repair which refers a good bond strength.

Flexural Behavior of Full-Scale Beams

Figure 7 shows the applied load versus midspan deflection as an average of two LVDTs. In general, the STF mixtures have shown a better flexural strength behavior than PPF mixtures. Naaman (*14*) reported that the flexural behavior is a function of fiber length and their bond strength. Therefore, the better behavior of the STF mixtures compared with the PP mixtures is mainly attributed to the longer fiber length as well as higher bond and stiffness. Also, the shape of the STF providing an additional increase in bond. The longer fibers will provide bridging stresses across the crack resulting in coalescence of the microcracks (*15*). Therefore, the toughness of the concrete mixtures was increased.

Table 5 summarizes the crack and ultimate loads for all the tested beams. The flexural cracking load ranged between 4.900 and 6.000 lb for repaired, and 4,600 and 4,700 lb for the reference control beams. Such an increase in the cracking load is due to using the FR-SCC that increased the tensile strength supported by the fibers in repaired zone. While, the ultimate load ranged between 15,492 and 26,200 lb for repaired, and 30,000 and 32,000 lb for the reference control beams. The

(a)

(b)

Figure 5. Failure modes observed in the compatibility test: (*a*) failure at the edge; (*b*) failure at the center.

Table 4. Results of Slant Shear Bond Strength Test

Age-day	Repair material	Slant shear bond strength, psi	Compressive strength ratio	Splitting tensile strength ratio	Failure mode
1	35SL25S	672	0.71	0.95	Substrate and repair
	35SL50S	734	0.62	1.20	Repair
	35SL15P	632	0.54	0.75	Repair
	35SL20P	651	0.39	0.61	Interface and repair
	10SF25S	309	0.94	1.11	Substrate and repair
	10SF50S	467	1.16	1.46	Substrate and repair
	10SF10P	587	1.00	1.02	Substrate and repair
	10SF15P	826	0.99	0.90	Substrate and repair
3	35SL25S	1857	0.86	0.84	Substrate and repair
	35SL50S	1964	0.76	1.10	Substrate and repair
	35SL15P	1456	0.76	0.85	Substrate and repair
	35SL20P	1458	0.54	0.75	Repair
	10SF25S	840	1.20	0.84	Substrate and repair
	10SF50S	958	1.43	1.10	Substrate and repair
	10SF10P	1229	1.06	0.85	Substrate and repair
	10SF15P	1953	0.98	0.75	Substrate and repair
28	35SL25S	3539	0.98	0.79	Substrate and repair
	35SL50S	3753	0.88	1.04	Substrate and repair
	35SL15P	2972	0.99	0.86	Substrate and repair
	35SL20P	3458	0.83	0.71	Substrate and repair
	10SF25S	2312	1.45	1.38	Substrate and repair
	10SF50S	3159	1.62	1.69	Substrate and repair
	10SF10P	3378	1.44	1.28	Substrate and repair
	10SF15P	3560	1.40	1.42	Substrate and repair

Figure 6. An example of the failure in both substrate and repair material.

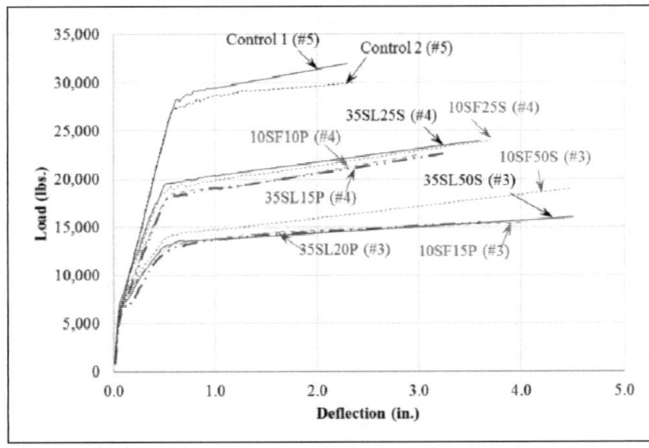

Figure 7. Load-deflection at mid-span of full depth for all the beams.

reinforcing steel bars continued in irreversible prolongation up to the ultimate load level. Depending on their tensile and bond strengths (pull out action), fibers help to resist tensile stress and arrest the enlargement of the cracks up to the ultimate load. Beams repaired with STF showed higher ultimate load than those repaired with PPF. For example, the ultimate load of the beam repaired with 10SF50S was 23% greater than the ultimate load of the beam repaired with 10SF15P. The longer length, crimped shape, and higher volume fraction of STF has mainly contributed to this performance. Swamy and Mangat (*16*) and Swamy and Jojaha (*17*) have proved that fiber type, volume, and geometry significantly affect the compaction and the ultimate load as a result.

Comparision of Cracking Load with the ACI 544

ACI 544 (*8*) considers the fibers in the concrete. The cracking load can be calculated by using Equation 1:

$$f_{cr} = 0.843 f_r . V_m + 425 RI \qquad ACI\ 544\ (8) \qquad (1)$$

$$V_m = 1 - V_f \qquad (2)$$

$$\mathrm{RI} = V_f\ L_f / d_f \qquad (3)$$

$$f_{cr} = \frac{M_{cr}}{I} Y \text{ therefore } \mathrm{M_{cr}} = \frac{I.f_{cr}}{\mathrm{Y}} \qquad (4)$$

$$\mathrm{I} = \frac{b.h^3}{12} (\text{for a rectangular section}) \qquad (5)$$

Table 5. Cracking and Ultimate Loads for the Tested Beams

Beam	Main reinforcement rebar #	Cracking load, lb	Increase in cracking load, %	Ultimate load, lb
Control 1	5	4700	–	32000
Control 2	5	4600	–	30000
35SL25S	4	5500	18.3	24000
35SL15P	4	4900	5.4	22700
10SF25S	4	5500	18.3	26200
10SF10P	4	5500	18.3	23000
35SL50S	3	5700	22.6	16000
35SL20P	3	5100	9.7	15492
10SF50S	3	6000	29.0	19000
10SF15P	3	5700	22.6	15500

Table 6. Results of Predicted Cracking Load by ACI 544 Code

	Pcr (lb)				
Beam	Exp.	Pred.	$Pcr_{exp}/Pcr_{Pred.}$	Mean	Mean
Control 1	4700	3360	1.40	1.38	1.12
Control 2	4600	3360	1.37		
35SL25S	5500	4036	1.36	1.05	
35SL15P	4900	5544	0.88		
10SF25S	5500	4974	1.11		
10SF10P	5500	5867	0.94		
35SL50S	5700	4377	1.3		
35SL20P	5100	6160	0.83		
10SF50S	6000	5312	1.13		
10SF15P	5700	6483	0.88		

$$Y = \frac{h}{2} \tag{6}$$

From the moment equilibrium ($\sum M = 0$):

$$M_{cr} = a\,\frac{p_{cr}}{2}\quad \text{, hence, } Pcr = \frac{2M_{cr}}{a} \tag{7}$$

where:

f_{cr}: first-crack composite strength, psi

f_r: stress in the matrix (modulus of rupture of the plain mortar or concrete, psi, according to ASTM C 78)

V_m: volume fraction of the matrix

V_f: volume fraction of the fibers

L_f / d_f: is the ratio of the length to diameter of the fibers (aspect ratio)

Table 6 shows the fcr values of the cracking load compared with the experimental values obtained from the load-deflection curve, or from direct observation. It can be seen that the model proposed by ACI 544 (*8*) underestimates the cracking load for both control and beams repaired with FR-SCC mixtures. The mean of the experimental to predicted cracking load (Pcr exp/Pcr pred) for all beams is 1.12. While the mean is 1.38 and 1.05 for control and repaired beams, respectively. Such a result is mainly attributed to the fact that ACI 544 code considers the effect of fibers.

Conclusion

The following are the main findings that can be drawn from the experimental results in this paper:

1. The incorporation of PPF fibers can largely affect SCC workability and placing ability in comparison with STF fibers. Therefore, the maximum fiber percent for the designed mixtures should not exceed 0.5% and 0.2% for STF and PPF, respectively.
2. The use of fiber in FR-SCC has improved all of the strength properties compared with the SCC mixtures.
3. A reduction of 37% in free drying shrinkage at 56 days was obtained when using 0.5% of steel fiber in the 35SL50S mixture compared with the 35SL mixture which is considered as the control mixture without fibers. Mixtures made with 35% SL exhibited lower shrinkage than those with 10% SF. By reducing the shrinkage crack size and distribution, the repaired structural element can be more durable.
4. In general, all optimized FR-SCC mixtures passed the compatibility test. The failure location was in the middle which indicates good compatibility.
5. Most of the FR-SCC mixtures showed good bonds with the substrate, which is an important factor for the successive repair. In general, the bond strength was found to increase with the increase in fiber content.
6. The repair with FR-SCC could be an efficient and viable option to repair the damaged beam, protect it from further loss, and to increase its cracking capacity. However, an additional effort should be made to recover the damaged reinforcement to maintain its original ultimate capacity.
7. The ACI 544 code first-crack composite strength provides safe prediction for both control and FR-SCC repaired beams.

Acknowledgments

The authors would like to acknowledge the financial support of RE-CAST (Research on Concrete Applications for Sustainable Transportation) Tier-1 University Transportation Center (UTC) at Missouri University of Science and Technology (Missouri S&T) at Rutgers University. The authors also would like to thank the Rutgers Infrastructure Monitoring and Evaluation (RIME) Group for their help in conducting the experimental work.

Author Contributions

The authors confirm contribution to the paper as follows: study conception and design: H.A., H.N., K.K.; data collection: H.A., H.N.;

analysis and interpretation of results: H.A., H.N., K.K.; draft manuscript preparation: H.A. All authors reviewed the results and approved the final version of the manuscript.

References

1. ASCE. *2017 Infrastructure Report Card*. ASCE, 2017. http://www.infrastructurereportcard.org. Accessed March 20, 2017.
2. Poston, R. W., K. Kesner, J. E. McDonald, A. M. Vaysburd, and P. H. Emmons. Selecting Durable Repair Materials: Performance Criteria Laboratory Results. *Concrete International*, Vol. 22, No. 11, 2000, pp. 21–29.
3. Abdulhameed, H. A. Evaluation of Pozzolanic Repair Materials by Different Test Methods. *Iraqi Journal for Mechanical and Materials Engineering*, Special Issue (A), 2010, pp. 1–14.
4. Hassan, M. S., A. S. Mansi, and H. A. Abdulhameed. Factors Affecting Compatibility between (S.B.R) Polymer Repair Materials and Concrete Substrate. *Engineering and Technology Journal*, Vol. 28, No. 14, 2010, pp. 4853–4865.
5. Mansi, A. S. Bond Strength Assessment for Different Types of Repair Materials. *Engineering and Technology Journal*, Vo. 28, No. 21, 2010, pp. 6325–6336.
6. Kassimi, F., A. K. El-Sayed, and K. H. Khayat. Performance of Fiber-Reinforced Self-Consolidating Concrete for Repair of Reinforced Concrete Beams. *ACI Structural Journal*, Vol. 111, No. 6, 2014, pp. 1277–1286.
7. Khayat, K. H., and W. J. Long. Shrinkage of Precast, Prestressed Self-Consolidating Concrete. *Materials Journal*, Vol. 107, No. 3, 2010, pp. 231–238.
8. ACI Committee 544. *Fiber Reinforced Concrete, ACI Manual of Concrete Practice, Part 6*. American Concrete Institute, Farmington Hills, 2011, pp. 11.
9. Brown, M. C., C. Ozyildirim, and W. L. Duke. *Investigation of Fiber-Reinforced Self-Consolidating Concrete*. No. FHWA/VTRC 10-R8. FHWA, Virginia Transportation Research Center, 2010.
10. Khayat, K. H., F. Kassimi, and P. Ghoddousi. Mixture Design and Testing of Fiber-Reinforced Self-Consolidating Concrete, *ACI Materials Journal*, Vol. 111, No. 2, 2014, pp. 143–152.
11. Czarnecki, L., A. Garbacz, P. Lukowski, and J. R. Clifton. *Polymer Composites for Repairing of Portland Cement Concrete: Compatibility Project*. NIST Interagency/Internal Report (NISTIR)-6394, US Department of Commerce, National Institute of Standards and Technology, Gaithersburg, MD, 1999.
12. Pattnaik, R. *Investigation into Compatibility between Repair Material and Substrate Concrete using Experimental and Finite Element Method*. PhD dissertation. Glenn Department of Civil Engineering, Clemson University, 2006.
13. Emmons, P. H. *Concrete Repair and Maintenance Illustrated*. R.S. Means Company, Inc., Kingston, Mass., 1993.
14. Naaman, A. E., and H. W. Reinhardt. *High Performance Fiber Reinforced Cement Composites 2 (HPFRCC2)*. Taylor & Francis, Ann Arbor, MI, 1996.
15. Shah, S. P., A. M. Brandt, C. Ouyang, R. Baggott, J. Eibl, M. A. Glinicki, H. Krenchel, A. Lambrechts, V. C. Li, B. Mobasher, and L. Taerwe. Toughness Characterization and Toughening Mechanisms. *Proc., RILEM*, Chapman & Hall, London, UK, 1996, pp. 193–228.
16. Swamy, R. N., and P. S. Mangat., Influence of Fiber Geometry on the Properties of Steel Fiber Reinforced Concrete. *Cement and Concrete Research*, Vol. 4, No. 3, 1974, pp. 451–465.
17. Swamy, R. N., and A. H. Jojaha. Impact Resistance of Steel Fiber Reinforced Lightweight Concrete. *Journal of Cement Composites and Lightweight Concrete*, Vol. 4, No. 4, 1982, pp. 209–220.

The Standing Committee on Basic Research and Emerging Technologies Related to Concrete (AFN10) peer-reviewed this paper (18-05860).

The contents of this report reflect the views of the authors, who are responsible for the facts and the accuracy of the information presented here.

Article

TRR
JOURNAL OF THE TRANSPORTATION RESEARCH BOARD

Transportation Research Record
2018, Vol. 2672(27) 56–64

DOI: 10.1177/0361198118777353
journals.sagepub.com/home/trr
SAGE

Bridge Substructure Repairs with Self-Consolidating Concrete and Galvanic Anodes

H. Celik Ozyildirim[1] and Stephen R. Sharp[1]

Abstract

Historically, the Virginia Department of Transportation (VDOT) has repaired chloride-contaminated reinforced concrete bridge substructure elements that contain vertical and overhead sections with either shotcrete or a conventional A3 (3,000 psi) or A4 (4,000 psi) concrete. This study investigated using self-consolidating concrete (SCC), which has a high flow rate, bonds well, has low permeability, and provides smooth surfaces, as another option. The study also explored the use of galvanic anodes to control corrosion activity in SCC repairs. In VDOT's Lynchburg and Staunton Districts, SCC repairs were made with and without the use of galvanic anodes. This provided a means for determining the benefit of using the anodes. The needed repair areas were determined by visual observation and sounding. After 7 years of service, SCC repair areas with and without anodes did not exhibit corrosion activity; small vertical cracks were evident in the SCC but did not affect performance. The anodes can provide protection to the steel immediately adjacent to the repair areas. However, unrepaired concrete areas away from the patched area with anodes now require additional repairs. SCC can be successfully placed; however, attention should be paid to form pressure and slump loss. Selection of repair areas should be based on corrosion-related measurements such as half-cell or chloride content, rather than sounding. Progression of corrosion demonstrates the necessity of removing all chloride-contaminated concrete not just adjacent to, but also away from the reinforcement, as anodes in the repair area will provide protection only in a narrow area around the patch.

In bridge substructure elements, leaking joints, shy cover depth, poor quality concrete, and lack of consolidation have commonly contributed to corrosion-related spalling and have led to costly repairs. Historically, repairs consist of removing deteriorated concrete and then placing either shotcrete or Virginia Department of Transportation (VDOT) A3 concrete with a minimum 28-day compressive strength requirement of 3,000 psi or VDOT A4 concrete with a minimum strength requirement of 4,000 psi. Shotcrete is projected under velocity to the surface, requiring special equipment and an experienced operator to ensure proper placement, making it a specialized and costly repair system (*1*). For VDOT A3 or A4 concrete, formwork is built that usually contains a larger volume than the area to be repaired to enable the placement and consolidation of concrete with conventional consistency. Therefore, the resulting repaired area often protrudes, does not match the geometry of the existing element, and is not aesthetically pleasing. In addition, improper consolidation can affect the quality of these repairs.

Self-consolidating concrete (SCC) is a comparatively new technology that provides very high workability (*2*). It has been used in Japan and Europe advantageously since the early 1990s (*3*). VDOT has used SCC successfully in beams, drilled shafts, and pier caps (*4–6*). SCC easily fills the congested spaces between the reinforcement and the formwork under the influence of its own mass and without any mechanical consolidation. SCC has good strength and reduced permeability because of a low water–cementitious material ratio (w/cm) to enable stability and the absence of large voids because of high workability. These characteristics prolong repair life by minimizing the intrusion of chlorides that initiate corrosion. Other advantages of SCC include smooth surfaces with a minimal finishing effort, a faster rate of concrete placement, and improved safety with no or limited mechanical vibration. The ability to test the concrete before placement also provides the opportunity for greater consistency in the quality of the repair concrete.

[1]Virginia Transportation Research Council, Charlottesville, VA

Corresponding Author:
Address correspondence to H. Celik Ozyildirim: celik@vdot.virginia.gov

The repair material (VDOT A3 or A4 concrete, shotcrete, or SCC) generally has a different chloride content and pH than the original concrete. These differences can lead to galvanic activity and ring corrosion, where the reinforcement adjacent to the repaired area corrodes leading to cracking and spalling of the concrete adjacent to the repair. To mitigate such corrosion, galvanic anodes containing zinc have been used in the repair area (*7*). In service, the zinc in galvanic anodes corrodes in place of the steel, thus protecting the steel from damage. It is also important to note that the galvanic anode system is self-contained and does not require an outside power source; however, the anodes have a limited life span, as they lose effectiveness once the zinc is depleted or has passivated.

Purpose and Scope

This study evaluated the use of SCC in repairing bridge substructure elements with and without galvanic anodes. The repair areas were selected based on visual observation and sounding. Bridge structures selected were in VDOT's Lynchburg District and Staunton District and are summarized in Table 1. In the bridges on Route 712 and Route 340, anodes were also placed and monitored. On the Route 340 Bridge, anodes were from three manufacturers, placed in different piers to determine the effectiveness of each. To determine the effectiveness of anodes, the current and voltage at the anode were measured.

SCC was tested for material properties at the fresh and hardened states. Assessment of the repairs was performed before, during, and after completion of repairs. The assessment involved an investigation of delamination by sounding with a hammer; corrosion activity was determined through half-cell potential and anode efficiency by current measurements. Corrosion-related data were collected for up to 7 years for the Route 712 Bridge and for 3.5 years for the Route 340 Bridge. Although the dates of the repairs varied, these repair sites were visited and visually reevaluated for several years, with the final evaluation done in spring 2017.

Methods

The study included field installations with supporting laboratory testing. Repairs in the field were made with SCC containing pozzolans, which reduce the permeability of concrete and improve its durability. Galvanic zinc anodes were placed in some of the repair areas to provide galvanic protection. These areas were compared with the control areas, which contained no anodes. Condition evaluations were performed routinely using visual examination, sounding, and current and half-cell potential measurements of the structures. The repair areas were selected based on visual observation and soundings.

The concretes in the substructures constructed prior to 1990 typically contain chlorides because of exposure to chloride spray from traffic, marine environments, or leaking joints; they rarely contain pozzolanic materials that would reduce permeability. However, SCC used in repairs does not contain chlorides from outside sources but does include a pozzolanic material for durability. Thus, the existing concrete and the SCC have different properties, including different chloride contents and pH values, which could lead to ring corrosion. However, SCC containing pozzolans has high resistivity that could reduce galvanic activity; anode manufacturers specify concretes with low resistivity for galvanic activity, which are generally concretes with high permeability. To mitigate this problem in repairs with low permeability concrete, some manufacturers provide embedding mortar that allows for low resistivity next to the anode. Embedding mortar was used in the Route 712 Bridge. The embedding mortar has high alkalinity, enabling an electrical circuit. In the Route 340 Bridge, calcium nitrite, an ionic material, was added to SCC to reduce the resistivity of concrete without compromising the low permeability characteristic of the SCC with pozzolans.

Field Installations

The SCC was sampled during placement, and the quality was assessed in the fresh and hardened states in accordance with the appropriate standard, shown in Table 2. For strength testing, concrete specimens were moist cured until testing. The permeability specimens were moist cured for 1 week at room temperature and then immersed in water at 100°F for 3 weeks before testing at 28 days. The drying shrinkage specimens were moist cured for 1 week and then dried at 50% relative humidity.

The corrosion activity of the SCC repair areas and surrounding concrete was monitored using half-cell

Table 1. Bridge Locations

Structure	Route	Location, over	District	Anodes
1	699	Hills Creek	Lynchburg	No
2	712	Route 29 Bypass	Lynchburg	Yes
3	I-81	Folly Mills Creek	Staunton	No
4	340	South Fork of the Shenandoah River	Staunton	Yes

Table 2. ASTM Tests for Fresh and Hardened Concrete Properties

Test	Specification
Fresh concrete	
Slump flow, T_{20}	C1611
Slump flow with J-ring	C1621
Air content	C173
Density	C138
Hardened concrete	
Compressive strength	C39
Elastic modulus	C469
Splitting tensile strength	C496
Modulus of rupture	C78
Permeability	C1202
Drying shrinkage	C157

Table 3. Mixture Proportions in lb/yd^3

Ingredient	Route 699 and 712[a]	I-81	Route 340
Portland cement, Type I/II	660	525	546
Fly ash, Class F	200	225	182
Coarse aggregate, No. 8	1,323	1,320	1,414
Fine aggregate, natural sand	1,220	1,404	1,364
Water	335	267	291
Design w/cm	0.39	0.36	0.40

Note: w/cm = water–cementitious material ratio.
[a]The mixture proportions for the Route 699 and 712 Bridges were the same; for the Route 699 Bridge, only slump flow and 28-day strength were determined.

potential measurements in accordance with ASTM C876. Some of the anodes also had external connections to enable current measurements.

Lynchburg District Bridges. The Route 699 Bridge repaired in 2010 did not include the installation of anodes and represented the first SCC use in a substructure by VDOT. Mixture proportions for both the Route 699 and Route 712 bridges in the Lynchburg District were the same and are summarized in Table 3. Commercially available air-entraining, regular water-reducing, and high-range water-reducing (HRWR) admixtures were added. Fresh concrete properties of the Route 712 bridge are presented in Table 4.

In the Route 699 Bridge, backwall and supporting buttresses with sharp corners were repaired with the SCC mixture. Concrete was placed by a bucket attached to a track hoe.

In the Route 712 Bridge, a pier cap and columns were repaired. The 49-ft-long cap was divided into three areas. One area had no anodes, but the other area had anodes that were embedded in two different ways: the middle area had anodes with embedding mortar, and the other area had anodes without the embedding mortar. The corroded steel was exposed, and about 1.5 in. was cleared behind the reinforcement. Anodes that were 1.25 in. thick were placed behind or next to the reinforcement along the perimeter against the existing concrete. A minimum clearance of 1 in. was planned around the anodes or the steel to ensure the flow of the SCC except when

Table 4. Fresh Concrete Properties of the Route 712 Bridge

Property	Age	Batch 1	Batch 2	Batch 3	Batch 4
		5/27/2010	6/9/2010	7/21/2010	8/18/2010
J-ring (in.)	a	23.2	28	27.5	30
T_{20} (s)[b]	a	3	2	2	1.5
Air content (%)	a	11.5	8.0	5.2	7.0
Density (lb/ft^3)	a	129.6	136.4	136.4	132.0
Concrete temp. (°F)	a	86	82	89	84
Compressive strength (psi)	1 day	2,800	–	2,780	2,050
	7 days	4,240	–	4,170	3,350
	28 days	5,430	–	5,130	4,090
Elastic modulus (×106 psi)	1 day	–	–	2.74	1.87
	7 days	3.5	–	–	2.95
	28 days	3.85	–	3.87	–
Splitting tensile strength (psi)	7 days	445	–	–	420
	28 days	530	–	–	515
Flexural strength (psi)	28 days	955	–	925	635
Permeability (coulombs)	28 days	844	–	1,906	2,055

Note: –= no data.
[a]Within 15 min.
[b]Time to reach a 20-in flow.

Figure 1. Placement of SCC from height to gain pressure (left), and manual placement method using container and bucket (right).

Figure 2. Small pump on deck of bridge (left); column being consolidated with internal vibrator (right).

embedding mortar was used, as it filled the area behind the anode.

Concrete was delivered using a funnel and flexible tube, as shown in Figure 1. At some intermediate points in the forms on the pier cap, buckets were used to place SCC, also shown in Figure 1. During placement, the wooden forms bulged and had to be strengthened. This caused delays, necessitating internal vibration to move the concrete.

Staunton District Bridges. There were two adjacent I-81 bridges, each with two piers. One pier cap and columns in the bridge carrying northbound traffic were repaired using SCC. The three other piers were repaired with shotcrete to serve as a comparison with the SCC. The mixture proportions for SCC used in the I-81 Bridge are also given in Table 3. Air-entraining, workability retaining, and HRWR admixtures were added. A small pump was used to place the SCC, as shown in Figure 2. Pumping allowed the SCC to flow fast, and the pump pressure enabled easy placement. However, because of the extended placement time, the SCC lost workability and internal vibration was used, as shown in Figure 2, to consolidate the concrete because it was losing workability.

For the SCC repairs for the Route 340 Bridge, anodes from three manufacturers were installed in two of the three piers repaired in an attempt to determine the effectiveness of these anodes in preventing corrosion. Concrete with the same mixture proportions including calcium nitrite was used in every location where anodes were placed for equal comparison; some locations were left with no anodes, as a control group. This mixture also contained an air-entraining, a retarding, and an HRWR admixture in addition to calcium nitrite. The pump malfunctioned the first day of placement, and small buckets were used to pour SCC manually from the top of the forms. The forms bulged in the first placement and had to be secured more firmly to the piers; the height of placement was limited to 8 ft.

Results and Discussion

Lynchburg District Bridges

Route 699 Bridge. SCC was tested at the jobsite and had a high slump flow (30 in.) without segregation. The 28-day strength was 7,610 psi. Placement was successful since the repaired backwall showed a very smooth finish and the buttresses had sharp corners.

Route 712 Bridge. In the Route 712 Bridge, half-cell potential measurements gathered prior to the SCC repairs indicated an average value along the top of the cap of −384 mV vs. saturated copper-copper sulfate electrode (CSE), with a standard deviation (SD) of 33 mV. At the bottom of the cap, values averaged −406 mV vs. CSE, with an SD of 69 mV. Although chloride-contaminated concrete was removed from the bottom of the cap, concrete in the top region of the cap was left in place even though the half-cell values indicated a greater than 90% probability of corrosion (based on values more negative than −350 mV vs. CSE). This was due to the fact that traditional sounding was used to determine where concrete should be removed rather than a direct assessment of the corrosion risk. The traditional sounding technique is limited to detecting delaminated concrete and will not detect concrete adjacent to corroding steel that has not yet delaminated. Selection of repair areas should be based on corrosion-related measurements such as half-cell or chloride content, rather than sounding.

The fresh and hardened concrete properties for the repair area along the bottom of the pier cap are shown in Table 4. The goal was to exceed 25 in. by the regular slump flow test without the J-ring. J-ring values tend to be lower than the regular slump flow because of the presence of the ring. In accordance with ASTM C1621, a difference between the regular slump flow and J-ring tests of less than 1 in. indicates good passing ability and a difference of more than 2 in. indicates poor passing ability. Batch 1 had delays in placement and exhibited a loss of workability; therefore, a workability retaining admixture was added to Batch 2, which was effective in retaining the slump flow. Buckets were used in Batch 4, which did not provide enough head pressure. When delays and slump loss occurred, internal vibrators were used.

When the concrete hardened and the forms were removed, large voids were observed on the top and bottom of the repair area placed on July 21, 2010. On the top, the ports to place the concrete were built too low, causing voids. On the bottom, the reinforcement was close to the surface with shy cover between the bottom of the form and the steel, which hindered the flow of SCC. Pouring with small buckets did not provide enough pressure to assist with flow, causing the large void at the bottom. These areas were repaired again. The new formwork provided a continuous open chute at the high end of the repair area, and a larger clearance between the bottom form and the steel was created by lowering the formwork from the initial level.

The concrete strengths for the Route 712 Bridge were satisfactory but highly variable, with the compressive strength ranging from 4,090 to 5,430 psi at 28 days. The strengths were lower than those for the Route 699 Bridge using the same mix design. The high variability was attributed to high variability in the air and water contents. SCC is sensitive to water content. The high water content in the Route 712 Bridge samples was also reflected in the high shrinkage values obtained for these concretes, as shown as in Figure 3 as an average of two batches.

Even though the shrinkage values were high, there were no wide cracks in the repaired areas after the completion of the repairs because of the heavy reinforcement and the limited size of the repair area. In a field survey of the repair area of the Route 712 Bridge on March 8, 2017, about 7 years after placement, vertical cracks ranging from a 1- to 4-ft spacing were observed, but they were tight. This survey did not find corrosion activity in the

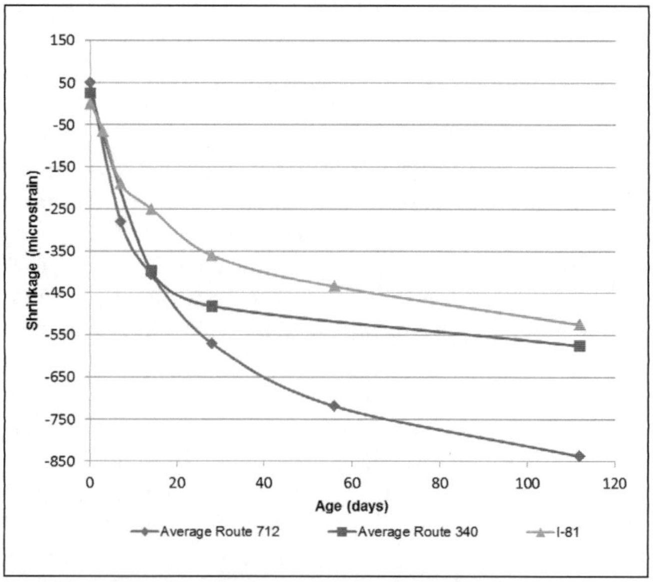

Figure 3. Shrinkage data for SCC in three bridges.

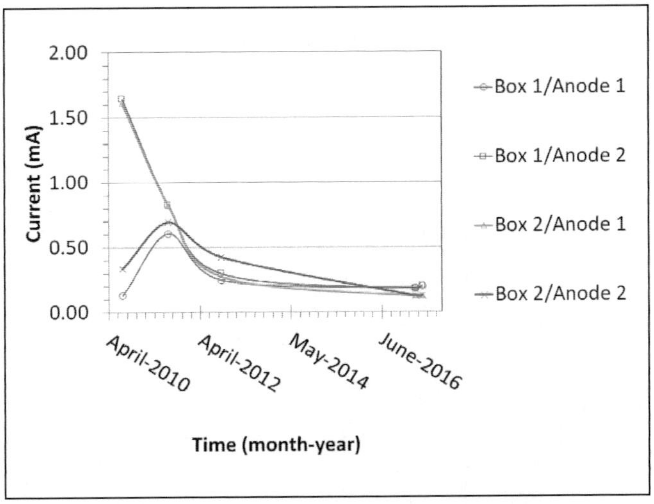

Figure 4. Current measured on Route 712 Bridge at four anode locations gathered over a 7-year period starting right after placement.

repaired area, and SCC was well bonded to the existing concrete, although it had vertical cracks. The anodes did provide protection to the steel adjacent to the repair area at a distance of about 6 in. away. Therefore, the anodes did provide a benefit for an anodic region adjacent to the cathodic patch, eliminating the halo effect. However, away from the limited protective zone, contaminated concrete showed distress and requires new repairs.

Measurements of current over time were gathered from four specially modified anodes. These anodes initially exhibited differing current outputs, which are shown in Figure 4. About one-third more charge was passed by the anodes that started at the higher current; however, after approximately 3 years, all of the anodes decreased to an equivalent level of current output, indicating that the effectiveness of the anodes had diminished.

There are efforts by the manufacturers to increase the amount of zinc in the anodes to extend the anode life and area of protection. However, the galvanic cathodic protection (CP) systems have a limited driving voltage and protection area as compared to an impressed current CP system as the galvanic potential difference is fixed by the voltage difference between the anode material selected and the steel being protected. Therefore, the CP options for repairing larger areas might be either to increase the number of galvanic anodes that are distributed in the old contaminated and the new concrete, or to select a CP system that has a greater driving voltage.

The presence of a wide distressed area away from the repaired areas is evident in the more recent half-cell data collected in May 2017, shown in Figure 5. There is a higher probability of corrosion in the upper region of the cap. The temperature in May started slightly cooler than in the summer months, reaching an approximate high of 65°F. Although this might have decreased the magnitude of the half-cell reading, it allowed for easier observation of cracks in the pier cap. Cracks were observed in the upper region of the pier cap, which was especially noticeable below the bridge seat. More negative half-cell potential readings were seen in regions with cracking that were not repaired. Although the lower portion of the structure where the repair was made several years ago is doing well and showing only tight cracks with no indication of corrosion, the older chloride-contaminated concrete that was not removed has continued to corrode and is now exhibiting damage.

Limited protection with the anodes placed in the repair areas raises questions about the cost-effectiveness

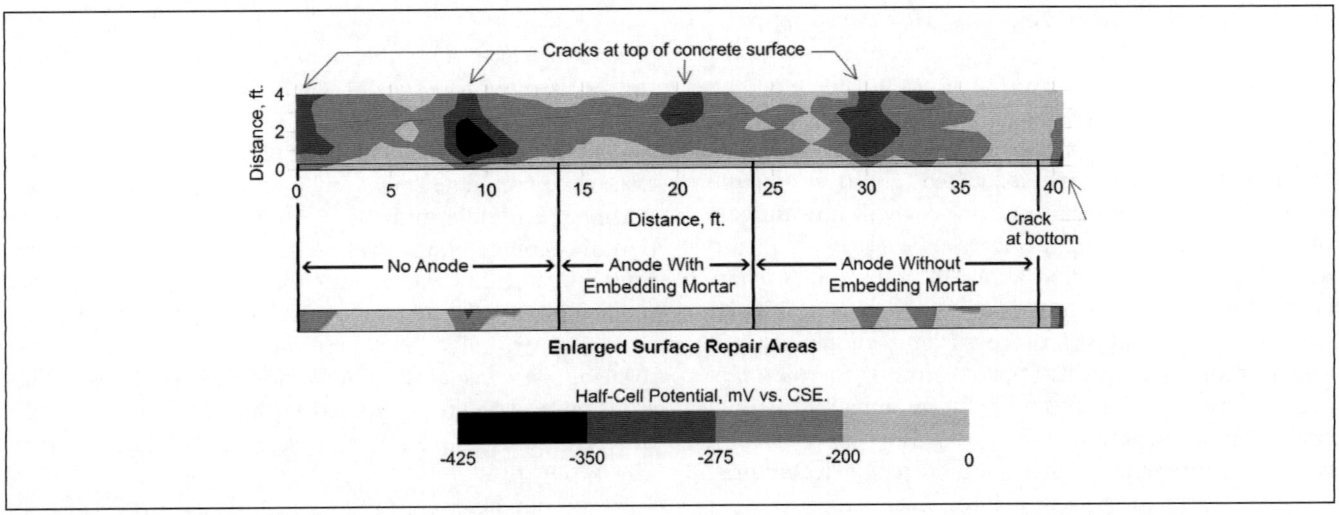

Figure 5. Recent half-cell potential map showing the crack location in the unrepaired area along the vertical face of the pier cap of the Route 712 Bridge, collected in May 2017.

Table 5. Fresh and Hardened Concrete Properties of the I-81 Bridge

Property	Age	B1 8/26/2011	B2 9/1/2011
Air (%)	[a]	5.8	4.6
T_{20} (s)	[a]	1.9	1.5
J-ring (in.)	[a]	23	31
Density (lb/ft^3)	[a]	138.0	143.2
Compressive strength (psi)	1 day	2,150	2,030
	7 days	3,620	3,820
	28 days	4,850	5,770
Splitting tensile strength (psi)	7 day	370	430
	28 days	490	460
Permeability (coulombs)	28 days	1,503	–

Note: –= no data.
[a]Fresh concrete.

Table 6. Fresh Concrete Properties of all the Batches for the Route 340 Bridge

Property	*N*	Min.	Max.	Avg.	SD
Slump flow (in.)	16	24.0	31.0	27.6	1.9
T_{20} (s)	16	0.5	3.9	1.8	0.9
Air (%)	14	2.4	10.5	6.9	2.3
Density (lb/ft^3)	16	129.6	146.8	137.4	4.5
Concrete temp. (°F)	14	67.0	84.0	77.9	5.4

Note: *N* = number of tests; Min. = minimum; Max. = maximum; Avg. = average; SD = standard deviation.

of requiring anodes in repairs when adjacent chloride-contaminated concrete is not removed. From what was observed at the Route 712 Bridge, if the area with chloride-contaminated concrete is limited and contained, galvanic anodes could be beneficial. In Figure 5, the area in the middle, which contains anodes with the embedding mortar, is showing the most favorable response with more uniform potential behavior. To the right of this middle area in Figure 5 is the location where anodes were placed without embedding mortar. This area is not performing as well as the middle area. However, it appears that the adjacent area to the repair with anodes is doing better than the adjacent area that does not have anodes even though none of the repair areas are exhibiting visible corrosion-related damage. Clearly, if the chloride-contaminated concrete area is located too far away from the patch repair area, galvanic anodes will not mitigate corrosion in the contaminated area unless they are placed there as well. In the region above the repair area where there are no anodes, the potential values are beginning to show a high probability of corrosion. Anodes do not provide protection in this upper region of the pier cap. This will result in a need for future repairs in those remote chloride-contaminated areas. For the protection of wider contaminated areas, other repair techniques, such as impressed current CP or even removal of a greater volume of chloride-contaminated concrete, should be considered.

Staunton District Bridges

I-81 Bridge. Two batches of concrete were tested during placement, and the properties are summarized in Table 5. The strength values varied but were satisfactory. The shrinkage values were lower than those for the Route 712 Bridge as expected, as the water contents and paste contents were lower. The repair with SCC compared with that with shotcrete resulted in a much smoother finish and was less noticeable aesthetically.

Route 340 Bridge. The minimum, maximum, and average fresh concrete properties of all the batches tested for the Route 340 Bridge are shown in Table 6. During the placement between 7/19/2013 and 8/23/2013, fresh concrete tests were conducted not only in the beginning but also during the middle and the end of placement to ensure that the concrete was still self-consolidating. A slump flow range of 25 to 30 in. was planned. The results indicated a slump flow range of 24.0 to 31.0 in. Concretes were very workable, and there was no objectionable segregation; the visual stability index was 1 or less. The addition of a retarding admixture had enabled workable concretes during placement. The specified air content was 4 to 9%.

Seven batches were tested for hardened properties, as indicated in Table 7. The air content and density for those batches are also included in the table. The 28-day

Table 7. Fresh and Hardened Concrete Properties of the Route 340 Bridge

		B1	B2	B3	B4	B5	B6	B7
Property	Age (day)	7/19/2013	7/23/2013	7/26/2013	8/2/2013	8/9/2013	8/16/2013	8/23/2013
Air (%)	[a]	7.9	10.5	2.4[b]	3.8[b]	4.3	6.9	–
Density (lb/ft^3)	[a]	136.8	133.6	146.8	143.2	142.4	134.8	129.6
Compressive strength (psi)	3	3,480	2,600	3,730	3,610	3,420	3,340	2,760
	7	3,600	5,250	3,960	4,410	5,040	3,590	3,630
	28	5,580	7,010	5,940	5,860	6,000	5,130	4,560
Permeability (coulombs)	28	1,358	384	978	1,026	1,596	1,259	3,221

Note: – = no data.
[a]Fresh concrete.
[b]More air-entraining admixture added, and air raised to approximately 9% for placement.

average strength was 5,726 psi and the lowest strength was 4,560 psi, indicating that the 3,000 psi specification limit was easily met. However, the variability was high, with an SD of 766 psi. The lowest strength was in the mixture with the lowest density. High variability was also observed in the permeability values, with an SD of 888 coulombs. The maximum requirement for the 3,000 psi concretes is 3,500 coulombs. Even the last batch with the lowest density used in a limited amount to complete the placement met the 3,500 coulomb requirement. This mixture is expected to have a high air content and w/cm to yield low density and strength compared with other batches.

The results for the Route 340 Bridge showed that concretes, in general, had high workability and were stable but had high variability in air content, density, strength, and permeability. The high variability was attributed to variations in air and water contents. Even though the results of SCC testing showed that the product was highly variable, concretes used in the structure are air entrained, meeting the specifications, and have satisfactory strength and permeability to provide the expected service life.

The recent condition survey was in mid-2017, at 3.5 years. At that time, the repairs showed no visible signs of corrosion. However, this is not sufficient time to discern differences between the areas with and without the galvanic zinc anodes (experimental and control sections), adjacent or near each other. However, corrosion activity and new repairs on these bridges will be tracked.

Conclusions

- Despite the earlier placement problems in the field, the repairs with SCC were completed successfully, and repair areas do not exhibit corrosion activity.
- SCC has high variability in air content, slump flow, strength, and permeability, indicating sensitivity to water; however, satisfactory values exceeding the minimum requirement were obtained.
- Form bulging with SCC can be prevented by securing forms firmly and restricting height of placement.
- SCC can lose workability during placement. Slump flow checks during placement would indicate if the SCC still had the desired workability. If the workability is reduced, mechanical vibration should be applied.
- If workability retaining admixtures or retarding admixtures are used in the mixture, workability of the SCC can be maintained during placement.
- SCC placement is facilitated by applying pressure through pumping or by pouring from a height. Placement by pouring from small buckets without any height does not provide the required flow through congested areas and narrow openings.
- After 7 years, chloride-contaminated concrete that was left in place during the original repair has initiated corrosion and requires repair.
- Selection of repair areas should be based on corrosion-related measurements such as half-cell or chloride content, rather than sounding.
- Seven years of experience has shown that anodes provide protection to the steel in a limited area adjacent to the repair but areas outside this limited region are not protected; corrosion is progressing in these unprotected areas and will require repair.

Acknowledgments

The authors thank the Virginia Transportation Research Council and the Federal Highway Administration for their support of this research.

Author Contributions

The authors confirm contribution to the paper as follows: study conception and design: H.C.O. and S.R.S.; data collection: H.C.O. and S.R.S.; analysis and interpretation of results:

H.C.O. and S.R.S.; draft manuscript preparation: H.C.O. and S.R.S. All authors reviewed the results and approved the final version of the manuscript.

References

1. American Shotcrete Association. *Shotcrete Contractor and Crew Qualifications*. 2014. https://www.shotcrete.org/media/pdf/ASAPositionPaper_Board_1.pdf. Accessed May 22, 2017.
2. American Concrete Institute. *237R-07: Self-Consolidating Concrete*. Farmington Hills, Mich., 2007.
3. Okamura, H., and M. Ouchi. Self-Compacting Concrete: Development, Present Use and Future. In *Self Compacting Concrete: Proceedings of the First International RILEM Symposium* (A. Skarendahl, and O. Petersson, eds.), RILEM Publications, Cachan Cedex, France, 1999.
4. Ozyildirim, C. *Bulb-T Beams with Self Consolidating Concrete on the Route 33 Bridge Over the Pamunkey River in Virginia*. Publication VTRC 09-R5. Virginia Transportation Research Council, Charlottesville, 2009.
5. Ozyildirim, C., and S. Sharp. *Preparation and Testing of Drilled Shafts with Self-Consolidating Concrete*. Publication VCTIR 12-R15. Virginia Center for Transportation Innovation and Research, Charlottesville, 2012.
6. Ozyildirim, C., and G. Moruza. *Bridge Beams and Pier Caps with Self-Consolidating Concrete at Nimmo Parkway*. Publication VTRC 16-R11. Virginia Transportation Research Council, Charlottesville, 2016.
7. Brown, M. C., and S. R. Sharp. *Effectiveness of Sacrificial Anodes in High-Resistivity Shotcrete Repairs*. Publication VTRC 05-R25. Virginia Transportation Research Council, Charlottesville, 2005.

The Standing Committee on Properties of Concrete (AFN20) peer-reviewed this paper (18-03056).

TRR
JOURNAL OF THE TRANSPORTATION RESEARCH BOARD

Article

Transportation Research Record
2018, Vol. 2672(27) 65–77

DOI: 10.1177/0361198118772951
journals.sagepub.com/home/trr
SAGE

Performance Evaluation of Concrete Pavement Slab Considering Creep Effect by Finite Element Analysis

Siming Liang[1], Ya Wei[1], Zehong Wu[1], and Will Hansen[2]

Abstract

Creep, as an intrinsic property of concrete material, will inevitably affect the performance of concrete pavement slabs in the field. However, the creep effect on the performances of concrete pavement slabs is far from being fully investigated. In this study, a test set-up is designed to measure the flexural creep of concrete beams exposed to both sealed and drying conditions. The measured flexural creep results are then modeled by the microprestress–solidification theory-based creep model which is incorporated into finite element analysis to evaluate numerically the creep effect on the moisture warping deformation, warping stress, and the total stress under traffic load in concrete slabs. Parameters including slab size, slab thickness, and subgrade modulus are considered. It is found that concrete creep has a significant effect on slab performance. Based on the measured creep properties in this study, the warping deformation of slabs can be reduced by 8–62%, and the warping stress and the total stress can be relaxed by at least 50%. Therefore, it is of importance to incorporate creep effect in analyzing warping deformation and stress generated in concrete pavement slabs. This study also provides a numerical methodology to the current performance evaluation of concrete slabs in the field.

Warping of Portland cement concrete pavement slabs caused by the changes of internal relative humidity of concrete is a well-recognized issue, which has a significant influence on the long-term performance of concrete pavements. Moisture is one of the main environmental factors that influence volumetric changes of cement concrete pavement (*1–3*). When the bottom part of a concrete slab has a greater moisture content than the top part, a negative moisture gradient will be induced. The bottom part of the concrete slab will experience more expansion (or less shrinkage) than the top part, resulting in an upward warping. Conversely, a positive moisture gradient will occur, and the top part of the slab will experience more expansion (or less shrinkage) than the bottom part, resulting in a downward warping (*4*). Usually, the warping deformation of concrete slab is restrained to a certain degree by the self-weight of slabs, the base layer, and the dowel bar, and so forth, and stress will generate in concrete slabs. When combined with traffic loads, accelerating fatigue failure might occur in concrete slabs, which will cause bottom-up transverse crack or top-down corner crack in concrete slabs.

Recognizing the importance of the warping behavior to the performance of concrete pavement, many researchers have investigated this issue and great progress has been made in this area (*1–3,5*). However, the moisture warping of concrete slabs has not been studied as fully as the temperature curling (*5–7*). Recently, Wei et al. found that the warping stress generated in concrete slabs under some severe drying conditions is comparable to the thermal stress caused by the daily temperature gradient (*8*). Therefore, it is of importance to consider moisture warping during performance evaluation.

Moisture warping is a process related to the formation of a moisture gradient along the concrete slab depth from external drying, and thus is a very slow phenomenon. For example, it is reported that the drying of a concrete specimen with dimension of 16 cm would last

[1]Department of Civil Engineering, Tsinghua University, Beijing, China
[2]Department of Civil and Environmental Engineering, University of Michigan, Ann Arbor, MI

Corresponding Author:
Address correspondence to Ya Wei: yawei@tsinghua.edu.cn

for about 10 years (9). During this long-term drying process, concrete creep would no doubt have a great impact on the warping deformation and the stress of concrete slab. Therefore, concrete creep needs to be considered during the performance evaluation of concrete pavement. However, investigations of the effect of concrete creep on the deformation and stress developments in pavement slabs are very limited. Yeon et al. and Lim et al. investigated the creep effect on warping stress in early-age concrete pavements and found that about half of the warping stress could be relaxed by concrete creep (*10,11*). It should be noted that neither of these research groups have conducted tests to measure the creep directly; the creep property was back-calculated from the measured strain history in the field (*10*) or the restrained ring test (*11*), and a more direct and accurate investigation on the effect of creep on the performance of concrete slab needs to be performed.

To this day the creep effect on pavement performance, including the warping deformation and total stress when the traffic loads are considered, has still not been studied comprehensively. Moreover, concrete creep is affected significantly by the moisture conditions in concrete (*12,13*). Any moisture variation in concrete slabs would enhance the creep behavior, which is believed to further affect the deformation and stress of concrete slabs.

This work intends to investigate the creep effect on the magnitude of warping and stress generated in concrete pavement slabs exposed to external drying, using finite element analysis. Flexural creep tests were conducted to obtain the concrete flexural creep properties under sealed and drying conditions, the microprestress–solidification (MPS) theory-based concrete creep model was used to characterize the flexural creep development. The creep effect on the warping deformation and stresses of concrete slabs was assessed by conducting sequential coupled moisture diffusion analysis and structural analysis. This study is expected to provide a numerical methodology for performance evaluation of concrete pavement considering creep effect.

Theoretical Basics

Moisture Field in Concrete Pavement Slab

To quantify the creep effect on the warping and stress in concrete slabs under drying condition, it is essential to determine the relative humidity (RH) distribution within the slabs. In concrete slab, two moisture-exchanging processes which affect the RH of concrete are considered in this study: (1) diffusion caused by external drying at the slab top surface and (2) self-desiccation caused by cement hydration within the entire concrete slab (*8*). The external drying process usually results in a non-uniform RH distribution along slab depth; however, the RH distribution resulting from the self-desiccation is uniform everywhere. In this study, Fick's law is employed to describe the moisture transport process within the concrete slab from external drying, which is expressed as:

$$\frac{\partial H}{\partial t} = \frac{\partial H_d}{\partial t} + \frac{\partial H_s}{\partial t} = \nabla \cdot [D(H) \cdot \nabla H] \qquad (1)$$

where, H_d is concrete RH as a result of external drying; H_s is concrete RH as a result of self-desiccation, which can be obtained from the measured RH of a sealed concrete specimen; $D(H)$ is the diffusion coefficient as a function of concrete RH.

Currently, there exist various methods to estimate the diffusion coefficient from local measurements of RH in concrete, the one proposed by Bažant and Najjar (*14*) is used in this study:

$$D(H) = D_0 \times \left\{ \alpha + \frac{1-\alpha}{1 + [(1-H)/(1-H_c)]^{\beta}} \right\} \qquad (2)$$

where, D_0 is the maximum value of $D(H)$ when H = 100%; α is the ratio of minimum to maximum values of $D(H)$; H_c is concrete RH when $D(H)$ = 0.5 D_0; β is the parameter characterizing the extent of the drop of $D(H)$ versus H curves.

The following boundary condition is applied on the top of the concrete slab:

$$-D(H)\frac{\partial H}{\partial n} = \zeta(H_{top} - H_{env}) \qquad (3)$$

where, ζ is the surface diffusion coefficient; $\frac{\partial H}{\partial n}$ is the relative humidity gradient in direction n of concrete drying surface; H_{top} is the concrete RH at the top surface; H_{env} is the environmental RH.

Because of the mathematical similarities between the moisture transport process and the heat transfer process, the moisture field in concrete slab can be easily calculated by incorporating the moisture transport equation (Equation 1) into the subroutine UMATHT in ABAQUS (*15*).

Finite Element Modeling of Concrete Slab Incorporating Creep Effect

Sequential coupled moisture diffusion analysis and structural analysis were performed to evaluate the performance of concrete slabs under environmental and traffic loads. Concrete pavement was modeled as a three-dimensional slab supported by the Winkler foundation that is represented by a series of nonlinear springs (as shown in Figure 1*a*). Three slab sizes of 5 m × 4 m,

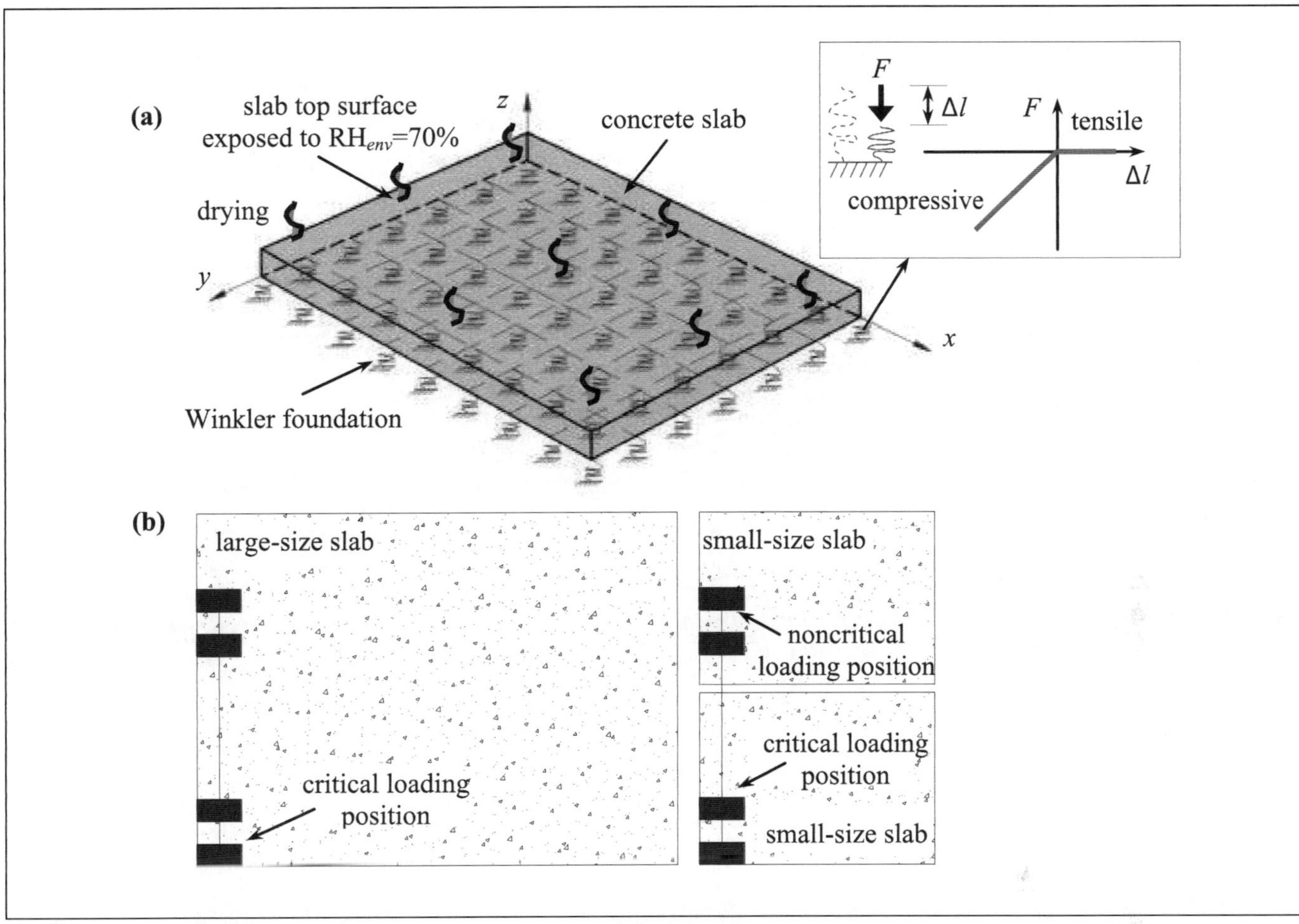

Figure 1. (*a*) Finite element model of concrete pavement slab for moisture diffusion analysis and structural analysis, and (*b*) critical loading position for large-size and small-size slab exposed to top surface drying.

2.5 m × 2.5 m, and 1 m × 1 m, three slab thicknesses of 17 cm, 20 cm, and 23 cm, and three subgrade moduli of 34 MPa/m, 68 MPa/m, and 136 MPa/m were analyzed in the finite element analysis (FEA). The sequential coupled FEA procedure is as follows. The moisture diffusion analysis is conducted to obtain the RH field of the concrete slab. During the moisture diffusion analysis, the top surface of the slab is assumed to be exposed to the environment with RH of 70% and the temperature is assumed to be constant. The duration of the analysis is about 28 days. DC3D20 element is used in the moisture diffusion analysis. Then the effect of RH on the performance of concrete pavement slabs is evaluated by structural analysis based on the RH results obtained in the moisture diffusion analysis. C3D20 element is used to mesh the concrete slab during structural analysis. At 28 days, a standard single axle load of 100 kN is applied at a slab corner to obtain the critical total stress of concrete slab (shown in Figure 1*b*), because this is the critical loading position when the concrete slab is exposed to top surface drying.

In this study, the MPS theory-based concrete creep model (*16,17*) is used to simulate the creep effect on slab deformation and stress. According to Bažant et al., the MPS theory is composed of two parts: the solidification theory part assumes that the creep aging mechanism is the result of volume fraction increase of the non-aging hydration products. The microprestress theory part is introduced to account for the temperature and RH effect. Changes in temperature and RH will create an imbalance in microprestress, which can help to reduce bonding between microstructure layers and thus increase creep rate.

The MPS based creep model is implemented in subroutines UMAT of ABAQUS. For step-by-step FEA, an incremental stress and strain relationship needs to be formulated. More details to obtain the incremental stress and incremental strain relationship can be found in the literature (*17,18*). In this study, it is assumed that the temperature during the simulation is constant. Therefore, one can obtain the three-dimensional stress–strain relationship for an isotropic material as:

$$\{\Delta\sigma_{n+1}\} = \bar{E}_{n+0.5}[Q]^{-1}\left[\{\Delta\varepsilon_{n+1}\} - \left\{\Delta\varepsilon_n^{ev''}\right\} - \left\{\Delta\varepsilon_n^{f''}\right\} - \{\Delta\varepsilon_n^{sh}\}\right] \quad (4)$$

where, $\{\bullet\}$represents a 6×1 tensor; $\bar{E}_{n+0.5} = 1/(q_1 + A_n^{ev} + A_n^f)$; $A_n^{ev} = \left[A_0 + \sum_1^N A_\mu(1-\lambda_{\mu,n})\right]/v_{e,m}$, $A_0 = 0.2794q_2$, $A_i = 0.2072\frac{(3\tau_i)^{0.1}}{1+(3\tau_i)^{0.1}}q_2$, $(i = 1,2,\ldots,10)$, $\tau_1 = 10^{-4}d$, $\tau_i = 10\tau_{i-1}$ $(i = 2,3,\ldots,10)$, $\lambda_{\mu,n} = (1-k_{\mu,n})/\Delta\gamma_{\mu,n}$, $\Delta\gamma_{\mu,n} = \psi_m\Delta t_n/\tau_\mu$, $\psi_m = \psi(t_{n+0.5}) = \alpha_H + (1-\alpha_H)H^2(t_{n+0.5})$, $\alpha_H = 0.1$, $k_{\mu,n} = \exp(-\Delta\gamma_{\mu,n})$; $v_{e,m}^{-1} = (\lambda_0/t_{n+0.5}^e)^m + \alpha_c$, $\lambda_0 = 1d$ $\alpha_c = q_3/q_2$, $m = 0.5$, $t_{n+0.5}^e = t_n^e + \frac{1}{2}\frac{1}{1+[\alpha_H - \alpha_H H(t_{n+0.5})]}\Delta t_n$; $A_n^f = \psi_m\Delta t_n/(2\eta_m)$, $\eta_m = \eta(S_{n+0.5}) = \frac{1}{2cS_{n+0.5}}$, $S_{n+0.5} = S_n + \frac{1}{2}\Delta S_n$, $\Delta S_n = -c_0\psi_{S,m}\Delta t_n S_m^2 + k_1\left|296\frac{\Delta H_n}{H_{n+0.5}}\right|$, $c_0 = \frac{2c}{q_4}$, $\psi_{S,m} = \alpha_S + (1-\alpha_S)H^2(t_{n+0.5})$, $\alpha_S = 0.1$; $\left\{\Delta\varepsilon_n^{ev''}\right\} = \left[\sum_1^N (1-k_{\mu,n})(A_\mu\{\sigma_n\} - \{\gamma_{\mu,n}\})\right]/v_{e,m}$; $\left\{\Delta\varepsilon_n^{f''}\right\} = \psi_m\{\sigma_n\}\Delta t_n/\eta_m$; $\{\Delta\varepsilon_n^{sh}\} = k_a(\Delta H_n, \Delta H_n, \Delta H_n, 0, 0, 0)^{\mathrm{T}}$, k_a is a coefficient that reflects the shrinkage strain increment caused by per RH increment;

$$[Q] = \begin{bmatrix} 1 & -\nu & -\nu & 0 & 0 & 0 \\ -\nu & 1 & -\nu & 0 & 0 & 0 \\ -\nu & -\nu & 1 & 0 & 0 & 0 \\ 0 & 0 & 0 & 2(1+\nu) & 0 & 0 \\ 0 & 0 & 0 & 0 & 2(1+\nu) & 0 \\ 0 & 0 & 0 & 0 & 0 & 2(1+\nu) \end{bmatrix}$$

.

Flexural Creep Test

To obtain the creep property of concrete, a flexural creep test was conducted on concrete beams. The beam size is 50 mm in height, 50 mm in width, and 1220 mm in length. The water/cement ratio of concrete used in this study was 0.30, the water content was 243 kg/m^3, and the mass ratio of fine aggregate to coarse aggregate was 0.50. Crushed limestone with a maximum aggregate size of 12.5 mm was used as coarse aggregate. According to the requirement of specimen size in ASTM C512 standard for concrete creep measurement in compression, the ratio of the diameter of specimens to the maximum aggregate size should be greater than 3. Therefore, the measured results in the study can reflect the homogenous creep property. The fine aggregate was natural sand with a fineness modulus of 2.56. Naphthalene-based high-performance superplasticizer was used to adjust the workability. The measured elastic modulus (tensile strength) values at 1 day, 3 days, 7 days, and 28 days by Φ100 mm $\times$ 400 mm specimen are 24.3 GPa (2.15 MPa), 28.0 GPa (2.63 MPa), 31.2 GPa (3.48 MPa), and 33.6 GPa (3.93 MPa). The measured elastic modulus will be used for evaluating the elastic behavior of concrete slabs by FEA. The measured tensile strength will be used to evaluate the stress condition of concrete beam in the flexural tests.

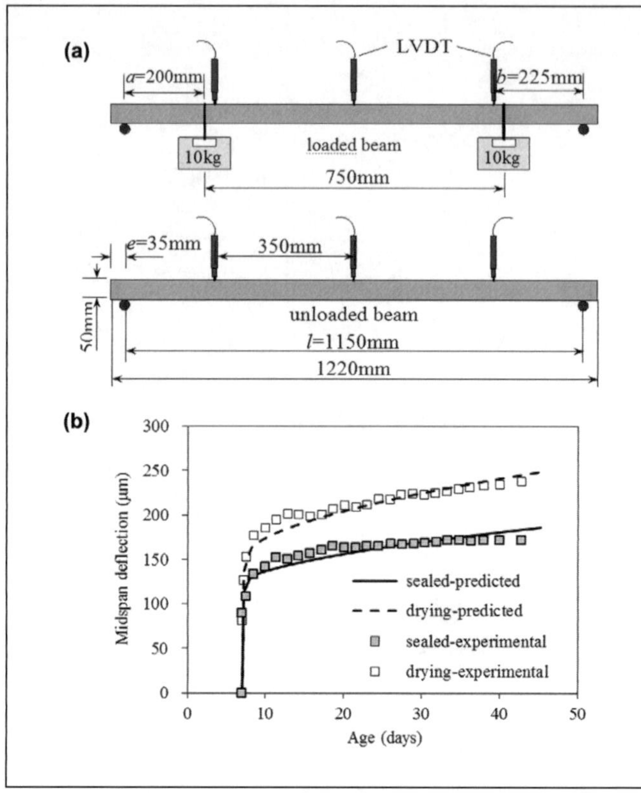

Figure 2. (*a*) Schematic illustration of four-point bending creep test, and (*b*) measured midspan deflection of concrete beam caused by external load under sealed and drying conditions.

The flexural creep test set-up was designed as a four-point bending configuration as shown in Figure 2*a*. The deflection of concrete beam was measured by using linear variable differential transformers with the precision of 1 μm. The measuring locations on concrete beam were at the midspan and the 350 mm from the midspan. During the test, the loaded beam was subject to two loads of 10 kg symmetrically about the midspan of the beam, and the loading positions were 200 mm away from the beam supports.

The creep properties of concrete beams under both sealed and drying condition were investigated. For the sealed case, the beams were sealed by three layers of self-adhesive aluminum foil in all faces. For the drying case, the top and bottom surfaces were exposed to the environment with RH = 50% and temperature of 23°C and other faces of the concrete beams were sealed to ensure a symmetric drying condition. Since the drying of all the concrete beams is symmetric and the shrinkage deformation is not reinforced, the shrinkage strain has a

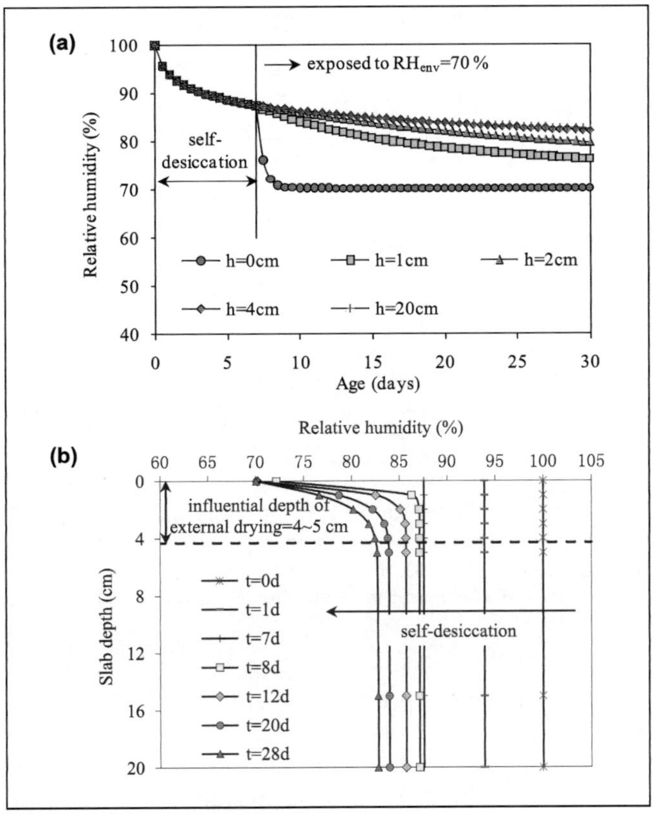

Figure 3. (*a*) Predicted relative humidity development at different depths of concrete slabs under 70% environment relative humidity, and (*b*) relative humidity distribution along the slab depth at different ages.

negligible effect on the measured deflection of concrete beams (*19*). All flexural creep tests started at 7 days and the measuring duration was about 30 days. The midspan deflection of the concrete beam caused by the external load can be obtained by deducting the measured deflection of the loaded beam by that of the unloaded beam, which is displayed in Figure 2*b*. It can be seen that the midspan deflection caused by external load under drying condition is greater than that under the sealed condition, which is consistent with the findings of Ghezal and Assaf (*19*).

To calibrate the MPS theory-based creep model by using the measured flexural creep results, the concrete beam was modeled and analyzed in ABAQUS according to the sequential coupled moisture diffusion analysis and structural analysis described in the previous section. The calibrated parameters of the MPS theory-based creep model are: $q_1 = 6.7 \times 10^{-6}$/MPa, $q_2 = 42.5 \times 10^{-6}$/MPa, $q_3 = 94.2 \times 10^{-6}$/MPa, $q_4 = 4.5 \times 10^{-6}$/MPa, $q_5 = 0.72$ day, $c = 7.0 \times 10^{-14}$ MPa^2/day, and $k_1 = 2.1$ MPa/K. The predicted midspan deflections of concrete beams are also plotted as lines in Figure 2*b*, showing that the MPS theory-based creep model can characterize the time-dependent deflection of concrete beam very well.

Results and Discussion

Development of Relative Humidity along Concrete Slab Depth

To calibrate the moisture diffusion model and to determine the shrinkage coefficient of concrete caused by the change of relative humidity, the shrinkage and internal RH of the 1000 mm × 100 mm × 38 mm concrete specimens under drying conditions were measured. More details about the measurement and the results of the shrinkage and the internal RH can be found in previous work (*20*). By matching the predicted RH with the measured one through FEA, the parameters of the moisture diffusion model can be calibrated. In addition, the shrinkage coefficient k_a is determined as 1.663×10^{-3} by linear regression analysis of the relation between the autogenous shrinkage and the internal RH of the sealed concrete specimen.

The calibrated moisture diffusion model was then used to simulate the distribution of RH along slab depth for concrete slabs with thickness of 17 cm, 20 cm, and 23 cm during the drying process. Figure 3*a* shows the variation of the RH with time at several typical slab depths. Figure 3*b* shows the RH development along the slab depth for the slab with thickness of 20 cm. It can be seen that the influential depth is about 4–5 cm from the drying surface at 28 days. The uniform self-desiccation only results in a straight RH line because of the uniform nature of self-desiccation.

Effect of Slab Size on Warping and Stress Considering Creep Effect

To evaluate the creep effect on the performances of concrete pavement in terms of warping deformation and stress generated in slabs, two types of analysis were conducted in ABAQUS for comparison: elastic analysis (with no creep), and viscoelastic analysis (with creep). Table 1 summarizes the calculated maximum warping deformation, the maximum warping stresses, and the maximum total stress when the traffic load was applied at the top surface of concrete slab at 28 days.

Slab size influences the warping deformation of concrete slab significantly, and the creep effect on the warping deformation is more pronounced in larger size slab. The 3D warping deformations of concrete slabs with different slab sizes are shown in Figure 4*a*. The warping deformation decreases with decreasing slab size in both the elastic analysis and the viscoelastic analysis. For example, the warping deformation of 1 m × 1 m × 20 cm slab is only about 8% and 19% of the warping deformation of 5 m × 4 m × 20 cm slab in the elastic analysis and the viscoelastic analysis, respectively. Moreover, it can be seen from Table 1 that the creep effect on the

Table 1. Summary of the Calculated Maximum Warping Deformation and Stresses in Concrete Slab at the Age of 28 Days

				RH_{env} = 70%		
Size (m × m)	Thickness (cm)	Subgrade modulus (MPa/m)	Analysis type	δ (mm)	S^w_x (MPa)	S^t_x (MPa)
1 × 1	20	68	Elastic	0.13	2.55	2.79
			Creep	0.12(8%)	0.72(72%)	1.01(64%)
2.5 × 2.5	20	68	Elastic	0.69	2.75	3.54
			Creep	0.54(22%)	0.92(67%)	1.78(50%)
5 × 4	17	68	Elastic	1.95	4.02	5.17
			Creep	0.75(62%)	1.34(67%)	2.59(50%)
	20	34	Elastic	1.66	3.51	4.87
			Creep	0.78(53%)	1.22(65%)	2.46(49%)
		68	Elastic	1.56	3.69	4.70
			Creep	0.64(59%)	1.25(66%)	2.25(52%)
		136	Elastic	1.48	3.85	4.53
			Creep	0.57(61%)	1.26(67%)	2.05(55%)
	23	68	Elastic	1.21	3.41	4.30
			Creep	0.53(56%)	1.16(66%)	1.99(54%)

Note: δ is the maximum warping deformation of concrete slab; S^w_x is the maximum warping stress in the *x* axis direction at the top surface of concrete slab; S^t_x is the maximum total stress in the *x* axis direction at the top surface of concrete slab. The value in bracket means the reduction of warping deformation or stress caused by creep behavior.

warping deformation is more significant in larger size slab. For example, the warping deformation can be reduced by from 8% to 59% as a result of the creep behavior when the slab size increases from 1 m × 1 m × 20 cm to 5 m × 4 m × 20 cm.

Slab size also influences the stress of concrete slab, and the creep effect on the stress is a bit more pronounced in smaller size slab. Figure 5*a* shows the maximum warping stress and the total stress within the concrete slab for different slab sizes. Both the warping stress and total stress decrease with decreasing slab size in the elastic analysis and the viscoelastic analysis. Besides, it can be seen from Table 1 that the creep property could relax 66–72% of the warping stress and 50–64% of the total stress, which is consistent with the findings of Yeon et al. and Lim et al. (*10,11*). Moreover, the creep effect on the reduction of the stress in 1 m × 1 m × 20 cm slab is slightly more pronounced than that in other larger size slabs.

The distributions of the warping stress and the total stress at the top surface of concrete slabs are also analyzed in this study. As shown in Figure 6, although the stress distributions at the top surface of concrete slabs are somewhat distinguished from each other, the maximum warping stress occurs at the center of the slab top surface and the maximum total stress occurs at the slab edge near the slab corner for all slab sizes.

Effect of Slab Thickness on Warping and Stress Considering Creep Effect

Slab thickness affects the warping deformation of concrete slabs, and the creep effect on the warping deformation is more pronounced in thinner slab. Figure 4*b* shows the 3D warping deformation of 5 m × 4 m slab with thickness of 17 cm, 20 cm, and 23 cm. It is observed that the warping deformation decreases with increasing slab thickness in both the elastic analysis and the viscoelastic analysis. For example, the warping deformation of 5 m × 4 m × 23 cm slab is only about 62% and 71% of the warping deformation of 5 m × 4 m × 17 cm slab in the elastic analysis and the viscoelastic analysis, respectively. The creep effect on the reduction of warping deformation is slightly more obvious in thinner slabs. It can be seen from Table 1 that the reduction of the warping deformation caused by concrete creep increases from 56% to 62% when the slab thickness decreases from 23 cm to 17 cm.

Slab thickness also affects the stress of concrete slab, and the creep effect on the stress is nearly the same for concrete slabs with different thicknesses. Figure 5*b* shows the maximum warping stress and the total stress in the concrete slabs with different thicknesses. Both the warping stress and the total stress decrease with increasing slab thickness in the elastic analysis and the viscoelastic analysis. However, slab thickness has less influence on the warping stress than the total stress. The reason lies in that slab thickness has a twofold effect on the warping stress; a thicker slab could enhance the moisture gradient, which would induce greater stress in concrete slabs. On the other hand, a thicker slab could increase the stiffness of concrete slab, which would reduce the stress in concrete slabs. The interaction between these two effects accounts for the phenomenon that the slab thickness has less influence on the warping stress than the total stress. It can also be seen from Table 1 that the creep property

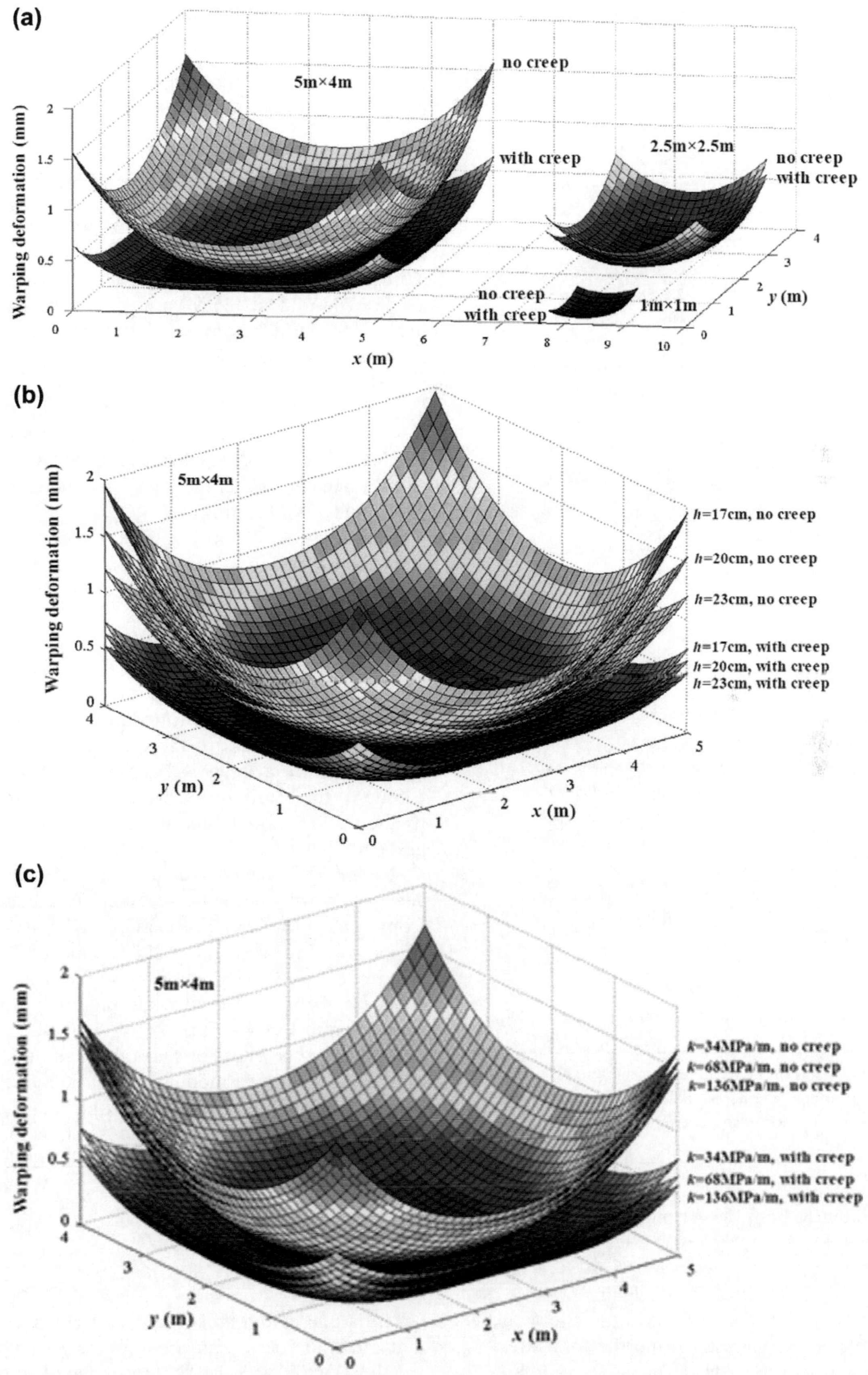

Figure 4. Comparison of slab warping deformation for (*a*) different slab sizes, (*b*) different slab thicknesses, and (*c*) supported by different foundations.

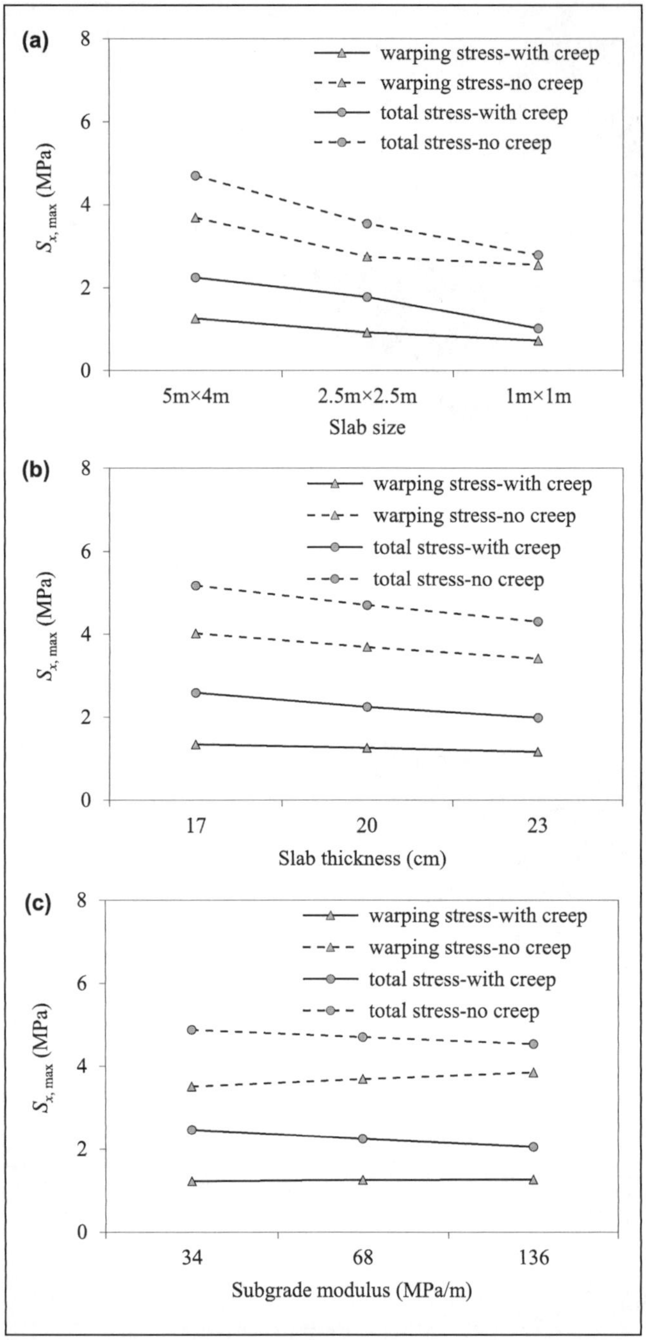

Figure 5. Comparison of the maximum warping stress and total stress at the top surface of concrete slab for (*a*) different slab sizes, (*b*) different slab thicknesses, and (c) supported by different subgrades at 28 days.

could relax about 66% of the warping stress and 52% of the total stress. However, the creep effect on the reduction of stress shows little difference in 5 m × 4 m slab with thickness of 17 cm, 20 cm, and 23 cm.

In addition, the slab thickness has little effect on the distribution of the warping stress and the total stress at the top surface of concrete slabs. As shown in Figure 7, the stress distributions of 5 m × 4 m slab with thickness of 17 cm and 23 cm are similar to each other.

Effect of Subgrade Modulus on Warping and Stress Considering Creep Effect

Subgrade modulus is another factor that affects the warping deformation, and the creep effect on the warping deformation is more significant in the case with larger subgrade modulus. Figure 4*c* displays the 3D warping deformation of 5 m × 4 m slab supported on Winkler foundation with subgrade modulus of 34 MPa/m, 68 MPa/m, and 136 MPa/m. It is seen that the warping deformation decreases with increasing subgrade modulus in the elastic analysis and the viscoelastic analysis. Moreover, the creep effect on the warping deformation is more significant in the case with larger subgrade modulus. For example, the warping deformation as a result of concrete creep can be reduced by from 53% to 61% when the subgrade modulus increases from 34 MPa/m to 136 MPa/m.

Subgrade modulus also affects the stress of concrete slab. Figure 5*c* shows the maximum warping stress and the total stress in concrete slabs supported on the foundation with different subgrade moduli. The warping stress increases with increasing subgrade modulus; however, the total stress decreases with increasing subgrade modulus. The reason is that more warping deformation is restrained by a stronger foundation, therefore a greater warping stress occurs in the case with larger subgrade modulus. Since a larger subgrade modulus would contribute to the bearing capacity of slab, less stress caused by the traffic load is found in the case with a larger subgrade modulus.

The creep effect on the reduction of stress does not vary much with the subgrade modulus. It can be seen from Table 1 that creep property could relax about 66% of the warping stress and 52% of the total stress. The creep effect on the reduction of stress of concrete slabs increases slightly with subgrade modulus ranging from 34 MPa/m to 136 MPa/m.

Similar to the effect of slab thickness on stress distribution, the subgrade modulus has little influence on the distribution of the warping stress and the total stress at the top surface of concrete slabs. As shown in Figure 8, the stress distributions of 5 m × 4 m slab in the case with subgrade modulus of 34 MPa/m and 136 MPa/m are similar to each other.

Conclusions

Creep is one of the most important concrete properties that affect the performance of concrete pavement slabs. In this study, the MPS theory-based creep model

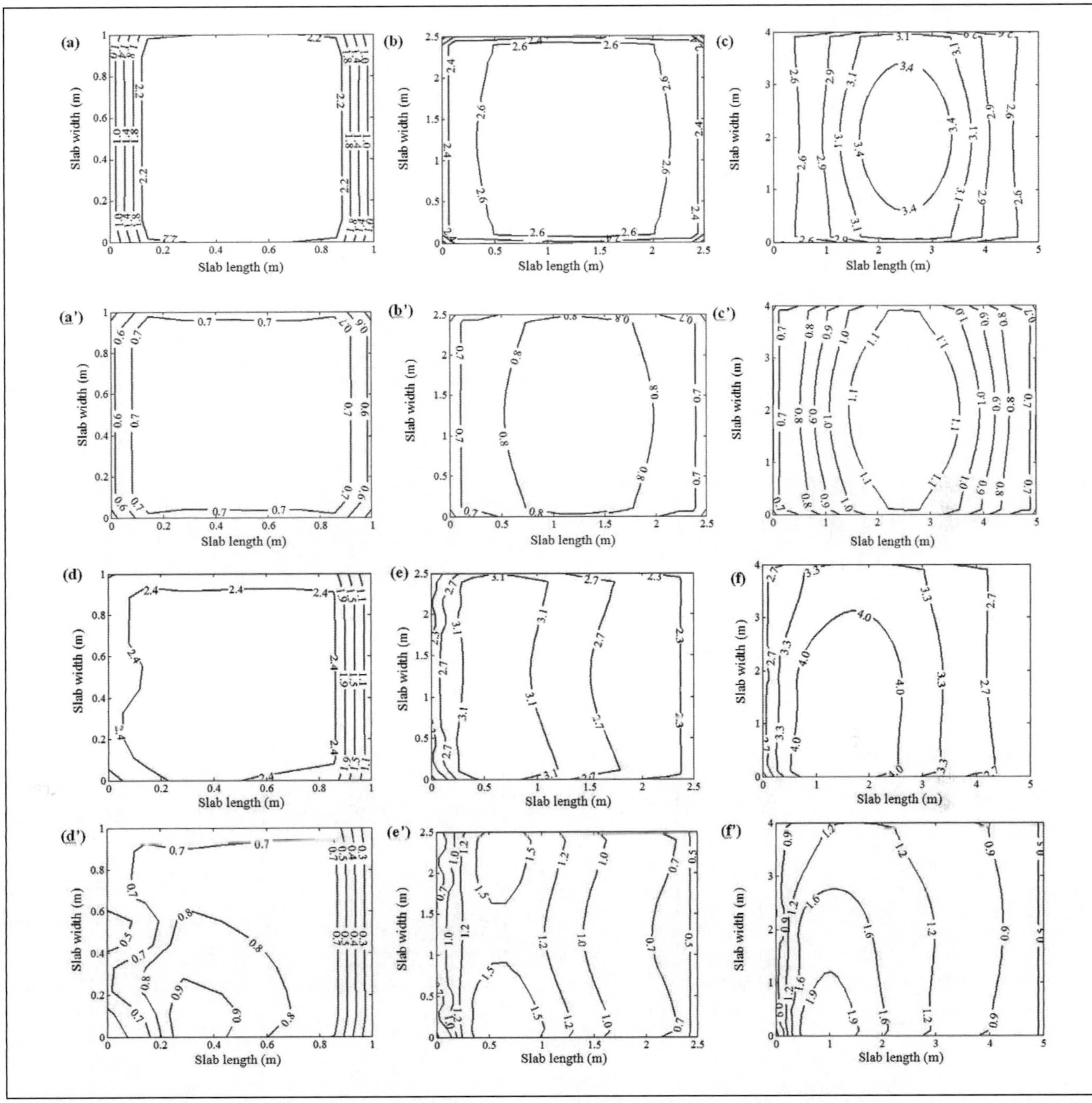

Figure 6. Warping stress (S_x) distribution at the slab top in concrete slab with size of (*a*) 1 m × 1 m × 20 cm, (*b*) 2.5 m × 2.5 m × 20 cm, and (*c*) 5 m × 4 m × 20 cm without creep effect, (a') 1 m × 1 m × 20 cm, (b') 2.5 m × 2.5 m × 20 cm, and (c') 5 m × 4 m × 20 cm by considering the creep effect, and total stress (S_x) distribution at slab top in concrete slab with size of (*d*) 1 m × 1 m × 20 cm, (e) 2.5 m × 2.5 m × 20 cm, and (*f*) 5 m × 4 m × 20 cm without creep effect, (d') 1 m × 1 m × 20 cm, (e') 2.5 m × 2.5 m × 20 cm, and (f') 5 m × 4 m × 20 cm, by considering creep effect. (subgrade modulus = 68MPa/m, unit of S_x: MPa).

calibrated by the measured flexural creep results of concrete beams was incorporated in the finite element model. The creep effect on the performance of concrete pavement slabs with different slab sizes, thickness, and subgrade modulus was evaluated by the sequential coupled moisture diffusion analysis and structural analysis in terms of moisture warping and stress. The major findings are:

Concrete creep has a significant effect on the warping deformation and stresses generated in concrete pavement

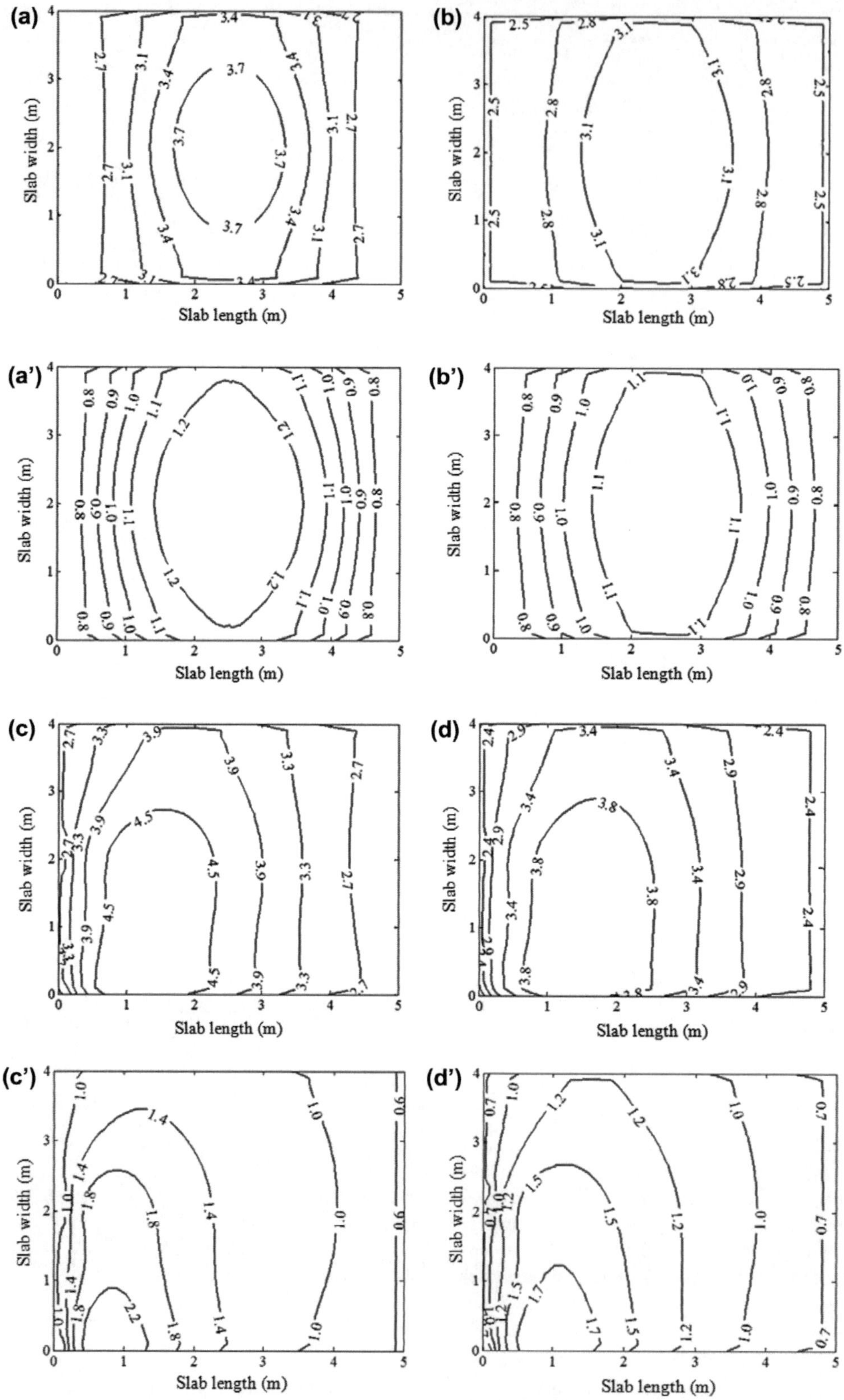

Figure 7. Warping stress (S_x) distribution at slab top in concrete slab with thickness of (*a*) 17 cm, and (*b*) 23 cm without creep effect, (a') 17 cm, and (b') 23 cm by considering creep effect and total stress (S_x) distribution at slab top in concrete slab with thickness of (*c*) 17 cm, and (*d*) 23 cm without creep effect, (c') 17 cm, and (d') 23 cm by considering creep effect. (slab size = 5 m × 4 m, subgrade modulus = 68 MPa/m, unit of S_x: MPa).

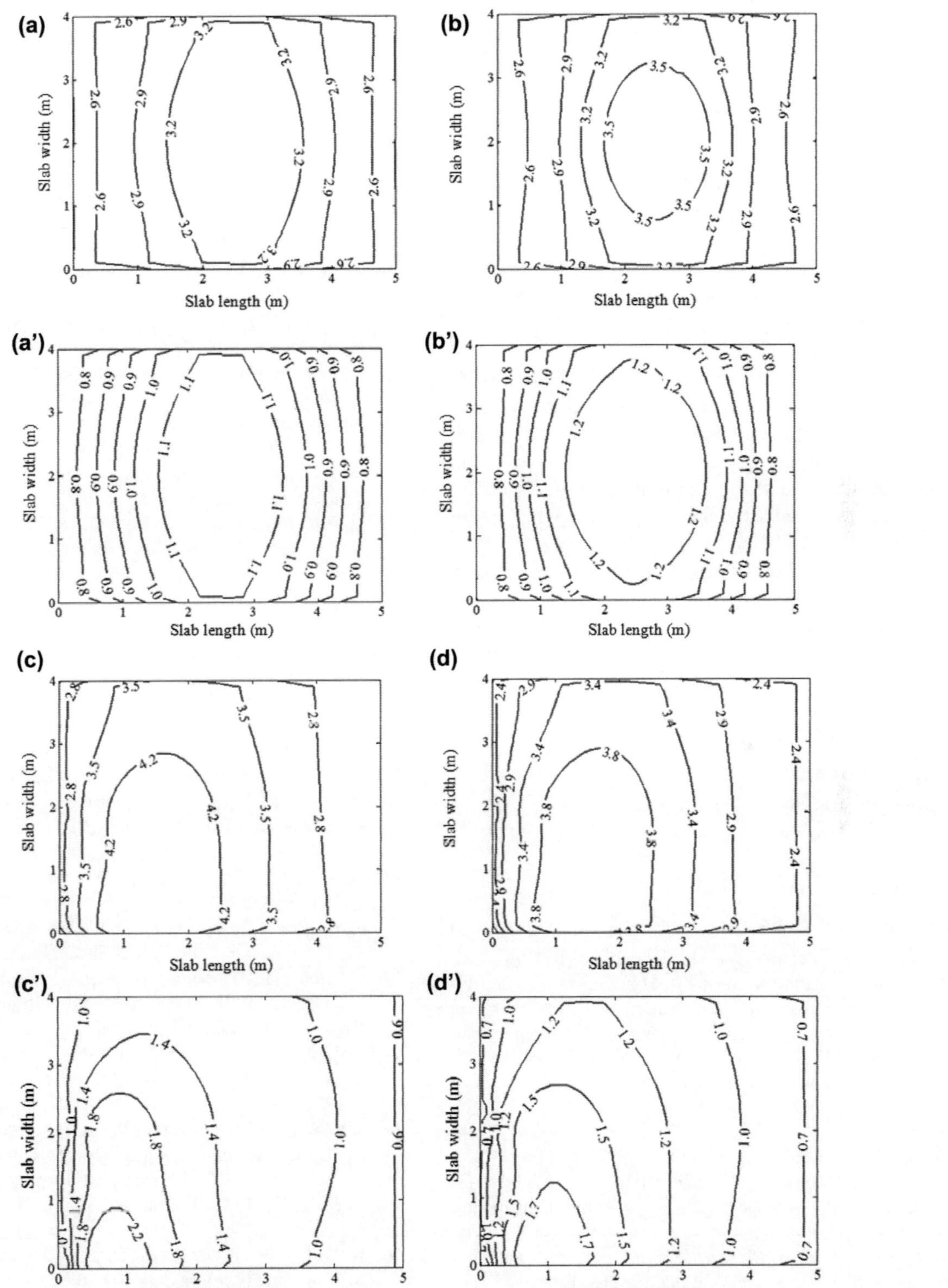

Figure 8. Warping stress (S_x) distribution at slab top in concrete slab supported on the foundation with subgrade modulus of (*a*) 34 MPa/m, and (*b*) 136 MPa/m without creep effect, (a') 34 MPa/m, and (b') 136 MPa/m by considering creep effect, and total stress (S_x) distribution at slab top in concrete slab supported on the foundation with subgrade modulus of (*c*) 34 MPa/m, and (*d*) 136 MPa/m without creep effect, and (c') 34 MPa/m, and (d') 136 MPa/m by considering creep effect. (slab size = 5 m × 4 m, thickness = 20 cm, unit of S_x: MPa).

slabs. Based on the flexural creep of concrete measured in this study, both the warping stress and the total stress can be relaxed or reduced by at least 50%. The warping deformation can be reduced by 8–62% from concrete creep. Therefore, it is of significance to consider the creep effect on the warping deformation and the stress of concrete slabs.

The creep effect on the warping deformation of concrete slab becomes more significant with increasing slab size and subgrade modulus. A thinner slab would also enhance the creep effect on the warping deformation. However, the creep effect on the stress of concrete slab does not vary much with varying slab size, thickness, and subgrade modulus.

Although concrete creep can reduce the warping deformation and stress of concrete slab, it does not alter the stress distribution pattern at slab top surface greatly. The maximum warping stress occurs at the center of the slab top surface, and the maximum total stress occurs near the slab edge when the traffic load is applied at the slab corner.

The results of this study prove the importance of considering concrete creep effect on the magnitudes of warping and stresses generated in concrete pavement slabs. The findings of this study are expected to provide a numerical methodology to the current performance evaluation of concrete pavement in the field.

Acknowledgments

The authors wish to thank National Natural Science Foundation of China under Grant No. 51578316 and 51778331 for the supports.

Author Contributions

The authors confirm contribution to the paper as follows: study conception and design: Ya Wei, Will Hansen; data collection: Siming Liang; analysis and interpretation of results: Ya Wei, Siming Liang, Zehong Wu; draft manuscript preparation: Siming Liang. All authors reviewed the results and approved the final version of the manuscript.

References

1. Jeong, J. H., and D. Zollinger. Early-age Curling and Warping Behavior: Insights from a Fully Instrumented Test-slab System. *Transportation Research Record: Journal of the Transportation Research Board*, 2004. 1896: 66–74.
2. Asbahan, R. E., and J. M. Vandenbossche. Effects of Temperature and Moisture Gradients on Slab Deformation for Jointed Plain Concrete Pavements. *Journal of Transportation Engineering*, Vol. 137, No. 8, 2011, pp. 563–570.
3. Wei, Y., S. Liang, and X. Gao. Numerical Evaluation of Moisture Warping and Stress in Concrete Pavement Slabs with Different Water-to-Cement Ratio and Thickness. *Journal of Engineering Mechanics*, Vol. 143, No. 2, 2016a, pp. 04016111.
4. Ceylan, H., R. F. Steffes, K. Gopalakrishnan, S. Kim, S. Yang, and K. Zhuang. Development and Evaluation of a Portable Device for Measuring Curling and Warping in Concrete Pavements. *InTrans Project Reports*, 2016, 176.
5. Bissonnette, B., E. K. Attiogbe, M. A. Miltenberger, and C. Fortin. Drying Shrinkage, Curling, and Joint Opening of Slabs-on-ground. *ACI Materials Journal*, Vol. 104, No. 3, 2007, pp. 259–267.
6. Wei, Y., and W. Hansen. Characterization of Moisture Transport and Its Effect on Deformations in Jointed Plain Concrete Pavement. *Transportation Research Record: Journal of the Transportation Research Board*, 2011. 2240: 9–15.
7. Hajibabaee, A., and M. T. Ley. Impact of Wet and Sealed Curing on Curling in Cement Paste Beams from Drying Shrinkage. *ACI Materials Journal*, Vol. 112, No. 1, 2015, pp. 79–84.
8. Wei, Y., X. Gao, and W. Hansen. Influential Depth by Water Absorption and Surface Drying in Concrete Slabs. *Transportation Research Record: Journal of the Transportation Research Board*, 2013. 2342: 76–82.
9. Acker, P., and F. J. Ulm. Creep and Shrinkage of Concrete: Physical Origins and Practical Measurements. *Nuclear Engineering & Design*, Vol. 203, No. 2, 2001, pp. 143–158.
10. Yeon, J. H., S. Choi, S. Ha, and M. C. Won. Effects of Creep and Built-in Curling on Stress Development of Portland Cement Concrete Pavement Under Environmental Loadings. *Journal of Transportation Engineering*, Vol. 139, No. 2, 2012, pp. 147–155.
11. Lim, J. S., S. H. Kim, and J. H. Jeong. Testing and Analysis of Viscoelastic Characteristics of Solidifying Concrete Pavement Slabs. *KSCE Journal of Civil Engineering*, Vol. 18, No. 4, 2014, pp. 1063–1071.
12. Hilaire, A., F. Benboudjema, A. Darquennes, Y. Berthaud, and G. Nahas. Modeling Basic Creep in Concrete at Early-age Under Compressive and Tensile Loading. *Nuclear Engineering and Design*, Vol. 269, 2014, pp. 222–230.
13. Cagnon, H., T. Vidal, A. Sellier, X. Bourbon, and G. Camps. Drying Creep in Cyclic Humidity Conditions. *Cement and Concrete Research*, Vol. 76, 2015, pp. 91–97.
14. Bažant, Z. P., and L. J. Najjar. Nonlinear Water Diffusion in Nonsaturated Concrete. *Materials and Structures*, Vol. 5, 1972, pp. 3–20.
15. Cui, W., W. Chen, and N. Wang. Thermo-hydro-mechanical Coupling Analysis of Early-age Concrete with Behavioral Changes Considered and Its Application. *China Civil Engineering Journal*, Vol. 48, No. 2, 2015, pp. 44–53. (in Chinese)
16. Bažant, Z. P., A. B. Hauggaard, S. Baweja, and F. J. Ulm. Microprestress-solidification Theory for Concrete Creep. I: Aging and Drying Effects. *Journal of Engineering Mechanics*, Vol. 123, No. 11, 1997, pp. 1188–1194.
17. Bažant, Z. P., G. Cusatis, and L. Cedolin. Temperature Effect on Concrete Creep Modeled by Microprestress-solidification Theory. *Journal of Engineering Mechanics*, Vol. 130, No. 6, 2004, pp. 691–699.

18. Wei, Y., W. Guo, and S. Liang. Microprestress-solidification Theory-based Tensile Creep Modeling of Early-age Concrete: Considering Temperature and Relative Humidity Effects. *Construction and Building Materials*, Vol. 127, 2016b, pp. 618–626.
19. Ghezal, A. F., and G. J. Assaf. Time-dependent Behavior of Self-consolidating Concrete Loaded at Early Age: Influence of Chemical Admixtures. *Journal of Materials in Civil Engineering*, Vol. 28, No. 1, 2015, pp. 04015066.
20. Wei, Y., W. Guo, and X. Zheng. Integrated Shrinkage, Relative Humidity, Strength Development, and Cracking Potential of Internally Cured Concrete Exposed to Different Drying Conditions. *Drying Technology*, Vol. 34, No. 7, 2016c, pp. 741–752.

The Standing Committee on Properties of Concrete (AFN20) peer-reviewed this paper (18-02781).

Article

TRR
JOURNAL OF THE TRANSPORTATION RESEARCH BOARD

Transportation Research Record
2018, Vol. 2672(27) 78–87

DOI: 10.1177/0361198118777606
journals.sagepub.com/home/trr
SAGE

Durable Concrete Overlays in Two Virginia Bridges

H. Celik Ozyildirim[1] and Harikrishnan Nair[1]

Abstract

The purpose of this study was to implement innovative concretes with low permeability and reduced cracking potential in overlays to reduce chloride infiltration into the bridge decks. Two parallel bridges on Route 64 over Dunlap Creek in Alleghany County, Virginia, were selected for this study. For low cracking potential, relatively low water contents, shrinkage reducing admixtures, and lightweight aggregates were used. For low permeability, concretes had supplementary cementitious material and relatively low water–cementitious material ratios. In the overlays, five different materials were used: latex-modified concrete with rapid set cement; silica fume concrete alone; and silica fume concrete with shrinkage reducing admixture, lightweight coarse aggregate, and partial lightweight fine aggregate. A compressive strength of 3,000 psi at 3 days was sought. The performance of the overlay concretes was observed after two to three winters. The overlays used in this study achieved the specified strength and low permeability. There were minimal tight cracks except for one section with the latex-modified concrete with rapid set cement in the left lane of the westbound bridge. The extensive cracks in that section were attributed to plastic shrinkage from adverse weather conditions at placement and the fact that a truck had caught fire in that lane. Silica fume concrete overlays with shrinkage reducing admixture, lightweight coarse aggregate, or lightweight fine aggregate are ready for implementation in the field for low cracking overlays.

Bridge decks subjected to distress because of corrosion, salt contamination, and wear can be rehabilitated by the use of overlays. Overlays are expected to resist the penetration of solutions and improve surface characteristics. Corrosion of the reinforcing steel is a common form of distress in bridge structures initiated and accelerated by the penetration of chlorides to the level of steel. Innovative concretes are needed that have low permeability and reduced cracking potential to reduce the chloride infiltration and extend the service life of the structures. Depending on the type of overlay selected, the expected lifetime can be extended 5–30 years (*1*). A survey for an NCHRP study has shown that states have been using asphalt overlays with a waterproofing membrane or concrete overlays such as high-performance concrete, concrete with low slump and a low water–cementitious material ratio (w/cm), silica fume concrete (SFC), fly ash concrete, polymer concrete, and latex-modified concrete (LMC) (*1*).

Since the late 1960s, the Virginia Department of Transportation (VDOT) has been investigating and using overlays that can protect decks and extend service life considerably (more than 25 years). Concretes that bond well and have low permeability such as SFCs, concretes with latex modifiers with Portland cement, or concretes with rapid set cement with normal weight aggregates are used extensively (*2–5*). Low permeability alone is not adequate; concretes that crack can facilitate the intrusion of solutions that cause deterioration. Thus, concretes that have low cracking potential and concretes with very tight cracks (less than 0.1 mm) are desired. Cracks less than 0.1 mm in width hinder the penetration of chlorides, which is the main source of distress (*6–8*).

To reduce the cracking potential in bridge decks, VDOT uses either normal weight concrete with low shrinkage or lightweight concrete; both have a maximum cementitious material content (*9*). If concrete with normal weight aggregate does not yield the low shrinkage of less than 0.035%, a shrinkage reducing admixture (SRA) is required to ensure the limit is met. Shrinkage is measured at 28 days of drying after 7 days of moist curing. Lightweight concretes (LWC) are not tested for shrinkage, as in a recent study on reducing cracks in bridge decks, LWC did not crack when shrinkage was above 0.035% and as high as 0.060% (*10*). The increased crack resistance in LWC is attributed to their lower elastic modulus, internal curing, and lower coefficient of thermal expansion compared with conventional concretes (*10, 11*). VDOT has been using LWC in decks with very good results for more than 30 years (*12*).

LMC and SFC overlays have low permeability and are widely used in Virginia. They are very thin (less than 2 inches

[1]Virginia Transportation Research Council, Charlottesville, VA

Corresponding Author:
Address correspondence to H. Celik Ozyildirim: celik@vdot.virginia.gov

thick) and are sensitive to cracking because of environmental conditions: loss of moisture resulting from high evaporation rates. An LMC overlay mixture must be mixed on-site in a mobile mixer and, thus, is more expensive than an SFC overlay mixture that can be mixed at any concrete plant or in a ready-mixed concrete truck and delivered to the site in the truck. For limited time applications, very-early-strength LMC with rapid set cement (RSLMC) can be used but must be produced in mobile mixers because of the fast setting time. In a research study on LMC overlays, Sprinkel showed that plastic shrinkage is more likely in LMC than in the regular VDOT A4 concrete used in bridge decks, which has a lower cementitious material content than LMC (*13*). In addition, LMC has a lower w/cm than VDOT A4 concrete, so less water is readily available to forestall plastic shrinkage. The study further concluded that if proper care is taken at the time of placement, plastic shrinkage cracking in LMC overlays is preventable (*13*). SFC also induces a tendency for plastic shrinkage cracking because of the fineness of the material (*14*). The fineness increases the water demand, leaving less free water to rise to the surface, and thus SFC is more susceptible to plastic shrinkage than conventional Portland cement concrete. Research indicated that SFC is vulnerable to plastic shrinkage just before initial set (*15*). Proper curing techniques are essential for the prevention of plastic shrinkage cracking. Both LMC and SFC overlays need to be covered with wet burlap and plastic promptly after screeding (*13*, *16*, *17*). According to Babaei and Hawkins, plastic shrinkage cracks in LMC overlays can be kept to a minimum by allowing a maximum evaporation rate of 0.73 $kg/m^2/hr$ (0.15 $lb/ft^2/hr$) while the overlay is placed (*16*). Kuhlmann (*18*) agreed with Babaei and Hawkins (*16*) on the maximum evaporation rate and recommended that the damp burlap and polyethylene film be placed immediately after the finishing operation. Luther recommended gently placing "dripping wet burlap" on each section of the overlay immediately after texturing (*19*).

LMC has been used on highway bridges for overlay rehabilitation for more than 40 years (*3*). Compared with concrete without latex, LMC is reported to be more resistant to intrusion of chloride ions; to have higher tensile, compressive, and flexural strength; and to have greater freeze–thaw resistance. The resistance to chloride intrusion is said to be attributable to the lower w/cm and a plastic film produced by the latex particles that prevents moisture loss (*3*). Sprinkel noted that high-early-strength concrete overlay mixtures with regular Type I/II cement, styrene butadiene latex, and 7% silica fume can be opened to traffic at 24 hours of age (*20*).

Whiting and Detwiler (*21*) evaluated bridge deck overlays with SFC and conventional low permeability dense concrete. Based on a field survey after 8 years of construction, the authors concluded that SFC appears to provide better protection against the ingress of chloride ions. The chloride profile showed lower chloride ion concentration in SFC compared with dense concrete at a given distance from the surface. Rehabilitating damaged deck slab concrete with an overlay system can increase the life of the reinforced concrete bridge deck and thus reduce life-cycle costs (*22*).

An SRA has a positive effect on lowering plastic and long-term drying shrinkage by reducing the surface tension of the pore water. Schemmel et al. reported that free shrinkage was reduced on the order of 50% with the use of an SRA (*23*). Maggenti et al. found that the use of an SRA in the concrete resulted in less shrinkage cracking in bridge decks (*24*). The use of prewetted lightweight aggregate can minimize the development of autogenous shrinkage and help to avoid early age-cracking (*10*). At the same time, its use has been shown to increase hydration; decrease shrinkage, porosity, and permeability; and increase compressive strength and mitigate shrinkage cracking through internal curing (*10*). Internal curing can also be achieved by replacing a portion of the fine aggregate with prewetted lightweight fine aggregate.

The main issues in overlays have been proper bonding, achievement of low permeability, and cracking. Another concern with overlays is the time it takes to open a bridge to traffic. Transportation agencies are facing a need to reduce overall construction time by modifying construction techniques, including materials used. Reducing construction time will reduce inconvenience to the traveling public and improve work zone safety. Proper immediate curing is required, which is not attained at all times. Therefore, more robust forgiving concretes would be desirable. As such, the successful use of concretes with SRAs and LWC in newly constructed bridge decks prompted VDOT to use these concretes in overlays.

Purpose and Scope

The purpose of this study was to implement innovative concretes with low permeability and reduced cracking potential in overlays to reduce chloride infiltration into bridge substructures and bridge decks. Two bridges on Route 64 over the Dunlap Creek in Alleghany County, Virginia, were selected for study. The westbound bridge is 578 feet long and has five spans, each 30 feet 8 inches wide; the three middle spans are each 115 feet 6 inches long; and the two end spans are each 115 feet 9 inches long . The eastbound bridge is 536 feet long and has five spans, each 30 feet 8 inches wide; the two end spans are each 101 feet 9 inches long, a second span is 101 feet 6 inches long, and the third and fourth spans are each 115 feet 6 inches long.

There are ten spans in total between the two bridges. The five overlay materials used in the study included (1) RSLMC and (2) SFC as control and (3) SFC with SRA, (4) SFC with lightweight coarse aggregate (SFC–LWCA), and (5) SFC with partial lightweight fine aggregate (SFC–PLWFA) for internal curing as shown in Figure 1. For low cracking potential, relatively low water contents, SRAs, and lightweight aggregates were used. This is the first use of LWC in a bridge

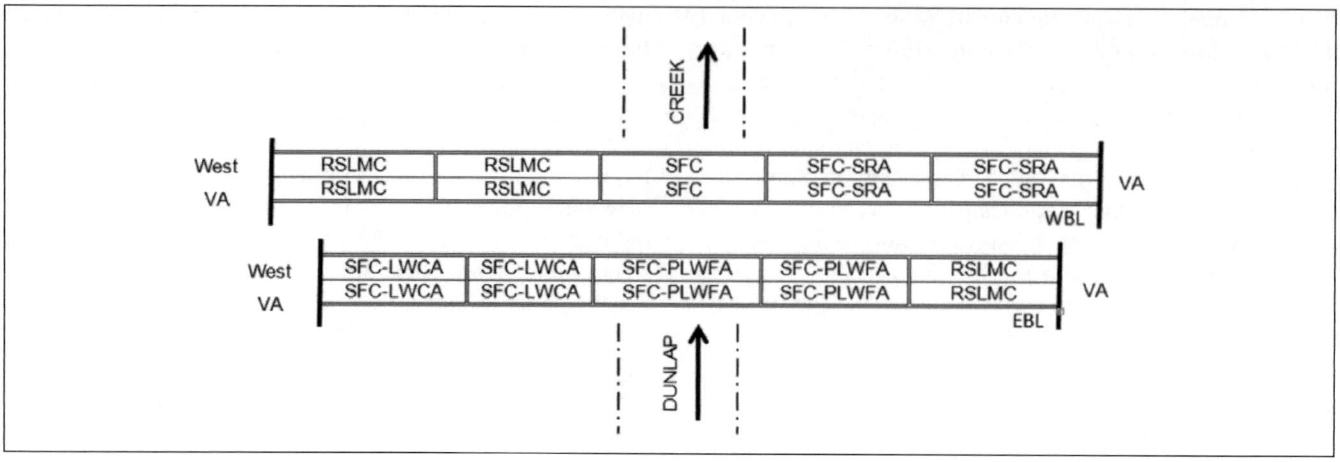

Figure 1. Overlay materials used in the two bridges: VA = Virginia; RSMLC = latex-modified concrete with rapid set cement; SFC = silica fume concrete; SFC–LWCA = SFC with lightweight coarse aggregate; SFC–PLWFA = SFC with partial lightweight fine aggregate; SFC–SRA = silica fume concrete with shrinkage reducing admixture; EBL = eastbound lane; WBL = westbound lane; FRC = fiber-reinforced concrete.

Table 1. Hardened Concrete Tests

Test	ASTM test	Specimen size (in)
Compressive strength	C39	4 × 8
Elastic modulus	C469	4 × 8
Splitting tensile strength	C496	4 × 8
Permeability[a]	C1202	2 × 4
Drying shrinkage[b]	C157	3 × 3 × 11
Freeze–thaw durability[c]	C666[c]	3 × 4 × 16

[a]Permeability specimens were moist cured 1 week at room temperature and 3 weeks at 100°F in lime saturated water.
[b]Drying shrinkage beams were moist cured for 7 days and then dried in the laboratory.
[c]C666 modified: specimens were moist cured for 2 weeks and then air dried for at least 1 week. The test water contained 2% NaCl. The freeze–thaw test was run for 300 cycles. The weight loss, durability factor, and surface rating were determined.

deck overlay in Virginia. For low permeability, concretes had supplementary cementitious material and relatively low w/cms.

Methods

In 2014, the westbound bridge was rehabilitated. The top surface of the deck was removed by hydrodemolition, and then the overlays were placed. In 2015, the eastbound bridge was rehabilitated in the same manner. To minimize traffic interruption, one lane was always kept open.

The overlays were required to achieve 3,000 psi compressive strength in 3 days before opening to traffic. Concretes were tested at the fresh state for air content (ASTM C231), slump (ASTM C143), density (ASTM C138), and temperature (ASTM C1604). Table 1 lists the ASTM test methods and specimen sizes for the tests on hardened concrete. For freeze–thaw durability a modified ASTM C666 test is used, which has a drying period before testing and has salt solution in the test water. The deck concretes and overlays are not exposed to cycles of freezing and thawing right after placement; usually a drying period occurs before the cycles, and in Virginia decks are subject to salt solution in deicing; therefore, salt solution is used in the test water.

About 1 3/4 inches of the surface of the old deck was removed using hydrodemolition. The concrete was mixed and delivered in ready-mixed concrete trucks except for RSLMC, for which a mobile mixer was used. Silica fume was added in dissolvable bags, which can leave bag pieces in concrete. In the early mixtures when the bags were added toward the end of the addition sequence, the bags did not dissolve and many pieces were removed by hand during placement. Later, the sequence of addition was moved to the beginning and the bags did dissolve, with a few pieces left in the concrete. The overlays were placed using a vibratory screed, which provided consolidation and leveling of the surface. The overlays were immediately covered with wet burlap and plastic.

In 2014, SFCs with and without SRA (similar mixture proportions but the SFC with SRA had 1.5 gallons of the admixture for 1 cubic yard) were used in the westbound bridge. The mixture proportions are shown in Table 2. SFC

Table 2. Mixture Proportions of Overlay Concretes (lb/yd^3)

Material	RSLMC	SFC	SFC–LWCA	SFC–PLWFA
Type I/II Portland cement	-	508	632	612
Rapid set cement	658	-	-	-
Class F fly ash	-	102		
Silica fume	-	25	48	46
LWCA, ½ in	-		790	–
Normal weight CA, No. 78	1,232	1481	–	1600
Sand	1,529	1451	1486	645
Latex	206	-	-	-
LWFA			–	495
Water	146	267	272	272
w/cm	0.39	0.44	0.40	0.41
AEA (oz/cwt)	-	0.4	0.4	0.4
Retarder (oz/cwt)	-	3	3	3
WRA (oz/cwt)	-	Varies	Varies	Varies
HRWRA (oz/cwt)	-	3.5	3.5	3.5
SRA (gal)	-	1.5	1.5	1.5
Air (%)	-	7	6.5	6.5

Note: RSLMC = latex-modified concrete with rapid set cement; SFC = silica fume concrete; SFC–LWCA = SFC with lightweight coarse aggregate; SFC–PLWFA = SFC with partial lightweight fine aggregate; CA = coarse aggregate; FA = fine aggregate; w/cm = water–cementitious material ratio; AEA = air-entraining admixture; WRA = water-reducing admixture; HRWRA = high-range water-reducing admixture; SRA = shrinkage reducing admixture; - = no data.

had ternary cementitious material composed of Portland cement, Class F fly ash, and silica fume. VDOT typically uses minimum 7% silica fume by weight of cementitious material in overlays (*9*). However, ternary systems are also permitted with reduced amounts of silica fume, which reduces the cracking potential. In the mixtures with lightweight aggregates, a binary system with 7% silica fume is used as lightweight aggregates are expected to reduce the cracking potential (*10*). A commercially available air-entraining admixture and a high-range water-reducing admixture were used in varying amounts to achieve the specified air content and workability. In 2015, the SFCs with lightweight coarse and fine aggregates were used in the eastbound lanes and the proportions are also given in Table 2. For SFC with lightweight fine aggregate, the lightweight aggregate added was approximately 53% of the volume of the fine aggregate.

The LWC mixtures with the lightweight coarse aggregate and natural normal weight sand had a maximum density of 120 lb/ft^3. In the mixtures with lightweight fine aggregate, there was no density requirement; the lightweight fine aggregate was used to provide internal curing.

Both RSLMC and SFC overlays were covered with wet burlap and plastic promptly after placement. Figure 2 shows overlay placement details.

Results and Discussion

The overlay concretes of the westbound bridge were placed 1 year before those of the eastbound bridge. The fresh and hardened concrete properties of these overlays were determined. The fresh concrete properties for the westbound bridge are summarized in Table 3, which also includes the weather data for some of the mixtures, and the properties for the eastbound bridge are summarized in Table 4. Workable concretes with satisfactory air contents were obtained. The average density of RSLMC was 141.8 lb/ft^3, SFC with and without SRA 142.5 lb/ft^3, SFC–LWCA 118.6 lb/ft^3, and SFC–PLWFA 136.0 lb/ft^3 as shown in Tables 3 and 4.

The hardened concrete properties are summarized in Table 5 for the westbound bridge and in Table 6 for the eastbound bridge. The specified compressive strength of 3,000 psi was achieved in 3 days. Concrete with lightweight aggregates achieved 3,000 psi in 1 day. This was attributed to a relatively low w/cm and internal curing. RSLMC attained 3,000 psi within 3.5 hours. The 28-day compressive and splitting tensile strengths were satisfactory, and permeability was low or very low as indicated by ASTM C1202. The elastic modulus was above 3 million psi at 28 days for all concretes.

The freeze–thaw durability results given in Tables 5 and 6 indicate that all concretes had a satisfactory durability factor. The acceptance criteria at 300 cycles are a weight loss of 7% or less, a durability factor of 60 or more, and a surface rating of 3 or less. The weight loss was within acceptance limits for the SFCs with normal weight and lightweight aggregates. The SFC with SRA and RSLMC had high weight loss, but the durability factors were satisfactory, indicating sound internal structure. All the permeability values were lower than the 1500 coulombs, as specified for overlays.

Figure 2. Overlay placement details at Dunlap Creek Bridge: (*a*) hydrodemolished surface, (*b*) placing overlay, (*c*) screeding and finishing, and (*d*) placing wet burlap and plastic.

Table 3. Fresh Concrete Properties of Overlays for the Westbound Bridge

	RSLMC1	RSLMC2	SRA1a	SRA1b	SRA2a	SRA2b	SFC1	SFC2
Test	9/12/14	10/8/14	9/15/14	9/16/14	10/10/14	10/10/14	9/15/14	10/10/14
Air (%)	5	4.2	6.1	7	5.2	6.8	6.4	6.4
Density (lb/ft^3)	142.0	141.6	-	143.2	-	141.2	142.8	142.8
Slump (in)	7.5	8.5	6	7	7	6	5.5	5
Concrete temp. (°F)	83	71	75	70	68	73	75	70
Air temp. (°F)	73.6	74	-	-	51	-	-	64.4
RH (%)	80.4	36	-	-	90	-	-	-
Wind (mph)	0	10	-	-	1	-	-	-

Note: RSMLC = latex-modified concrete with rapid set cement; SRA = shrinkage reducing admixture; SF = silica fume; RH = relative humidity. SRA was added to SFC.

SFC–LWCA B2 had higher coulombs than the SFC–PLWFA, which can be attributed to the former's higher w/cm than the design value, which is also shown by lower compressive strength in that mixture.

The shrinkage data are displayed in Figure 3 for both overlays. RSLMC had the lowest shrinkage values. SFC with SRA and SFC with lightweight coarse aggregate met the VDOT specification of 0.035% shrinkage at 28 days for

Table 4. Fresh Concrete Properties of Overlays for the Eastbound Bridge

	SFC–LWCA1		SFC–LWCA2		SFC–PLWFA1	SFC–PLWFA2		RSLMC1	RSLMC2
	7/20/15		8/3/15		7/22/15	8/4/15		7/23/15	8/5/15
Test	B1	B2	B1	B2	-	B1	B2	-	-
Air (%)	7.5	6	7.7	1	8.4	6	5.5	4.5	3.25
Density (lb/ft^3)	120.4	-	-	116.8	133.2	136.8	138.0	-	-
Slump (in)	5	5	2.25	3.75	2.5	3.5	3	4.5	5
Concrete temp. (°F)	82	-	74	75	-	85	85	83	83
Air temp. (°F)	-	-	-	-	-	-	-	77	-

Note: LWC = concrete with lightweight coarse aggregate; LWF = concrete with lightweight fine aggregate; RSMLC = latex-modified concrete with rapid set cement; SFC–LWCA = silica fume concrete with lightweight coarse aggregate; SFC–PLWFA = silica fume concrete with partial lightweight fine aggregate; B = batch; - = no data.

Table 5. Hardened Concrete Properties of Overlays for the Westbound Bridge

		RSLMC1	RSLMC2	SRA1a	SRA2a	SFC1	SFC2
Test	Age (days)	9/12/14	10/8/14	9/15/14	10/10/14	9/15/14	10/10/14
Compressive strength (psi)	1	4000	4710	2030	-	2120	-
	3	5620	5240	3040	2980	3110	2810
	7	5850	5605	-	-	3800	-
	28	6740	6180	5300	4770	5110	5260
E × 10^6 psi	28	3.52 (3 d)	3.69	3.82	2.70 (3 d)	3.71	2.79 (3 d)
Splitting tensile strength (psi)	28	690	580	500	530	500	555
Permeability (C)	28	348	292	638	905	507	841
Freeze–thaw durability at 300 cycles							
WL (%)		22.6	21.8	-	22.7	0.8	7.3
DF		75	81	-	87	102	105
SR		2.5	2.8	-	3.6	0.7	2.0

Note: RSMLC = latex-modified concrete with rapid set cement; SRA = shrinkage reducing admixture; SF = silica fume; E = modulus of elasticity, WL = weight loss; DF = durability factor; SR = surface rating.
SRA was added to SFC.

low cracking bridge decks. Babaei and Fouladgar recommended a maximum shrinkage of 0.07% at 4 months (*25*). SFC with SRA and SFC with lightweight aggregates also met this limit. SFC had shrinkage values above the limits set by VDOT (*9*) or Babaei and Fouladgar (*25*).

Surveys

Original Survey. Bridge decks were surveyed right after being covered by burlap and plastic for 1 week. The overlays did not have any visible cracks.

Second Survey. A later survey was made when the westbound bridge overlays (RSLMC and SFC with and without SRA) were 1 year old and the eastbound bridge overlays (SFC with lightweight coarse aggregate and lightweight fine aggregate) were 3–4 months old.

SFC with and without SRA or including lightweight aggregates exhibited no cracks or very tight cracks. SFC with SRA and LWCAs had low shrinkage (<0.035% at 28 days) and are expected to have minimal cracking. However, SFC and SFC–PLWFA had higher shrinkage than desired but still had tight cracks. This was attributed to good curing practices and favorable weather conditions. Other factors effective in controlling cracking in these concretes are: 1. SFC without the lightweight aggregates had ternary cementitious materials, which has a reduced cracking potential. 2. Lightweight aggregates are expected to reduce cracking potential in overlays mainly because of internal curing. 3. In this study all concretes had similar and low elastic moduli, as shown in Tables 5 and 6. Such low moduli would help in controlling cracking, as for a given deformation in the concrete the stresses are lower with low moduli. RSLMC performed well, as did the other overlays except that the left

Table 6. Hardened Concrete Properties of Overlays for the Eastbound Bridge

		SFC–LWCA1		SFC–LWCA2		SFC–PLWFA1	SFC–PLWFA2		RSLMC1	RSLMC2
		7/20/15		8/3/15		7/22/15	8/4/15		7/23/15	8/5/15
Test	Age (days)	B1	B2	B1	B2	B1	B1	B2	-	-
Compressive strength (psi)	1	-	-	3570	2800	3100	2760	3030	3690	(3.5 hr) 3020
	3	4470	4450	3740	3360	-	-	3760	-	(6 hr) 2950
	7	5240	5160	4580	3580	4800	4320	4740	4660	(8 hr) 3460
	28	6550	6550	5540	5070	6250	6030	6310	5280	5270
E × 10^6 psi	28	3.64	3.48	3.08	3.84	3.42	3.60	3.58	3.47	-
Splitting tensile strength (psi)	28	490	560	420	410	625	640	630	475	515
Permeability (C)	28	-	-	1359	1303	428	341	158	-	188
Freeze–thaw durability at 300 cycles										
WL (%)		0.1	0.5	1.6	0.1	0	2.3	4.6	-	-
DF		104	104	93	102	98	98	97	-	-
SR		0.2	0.6	0.7	0.2	0.2	0.6	0.8	-	-

Note: LWC = concrete with lightweight coarse aggregate; LWF = concrete with lightweight fine aggregate; RSMLC = latex-modified concrete with rapid set cement; SFC–LWCA = silica fume concrete with lightweight coarse aggregate; SFC–PLWFA = silica fume concrete with partial lightweight fine aggregate; B = batch; E = modulus of elasticity; WL = weight loss; DF = durability factor; SR = surface rating; - = no data.

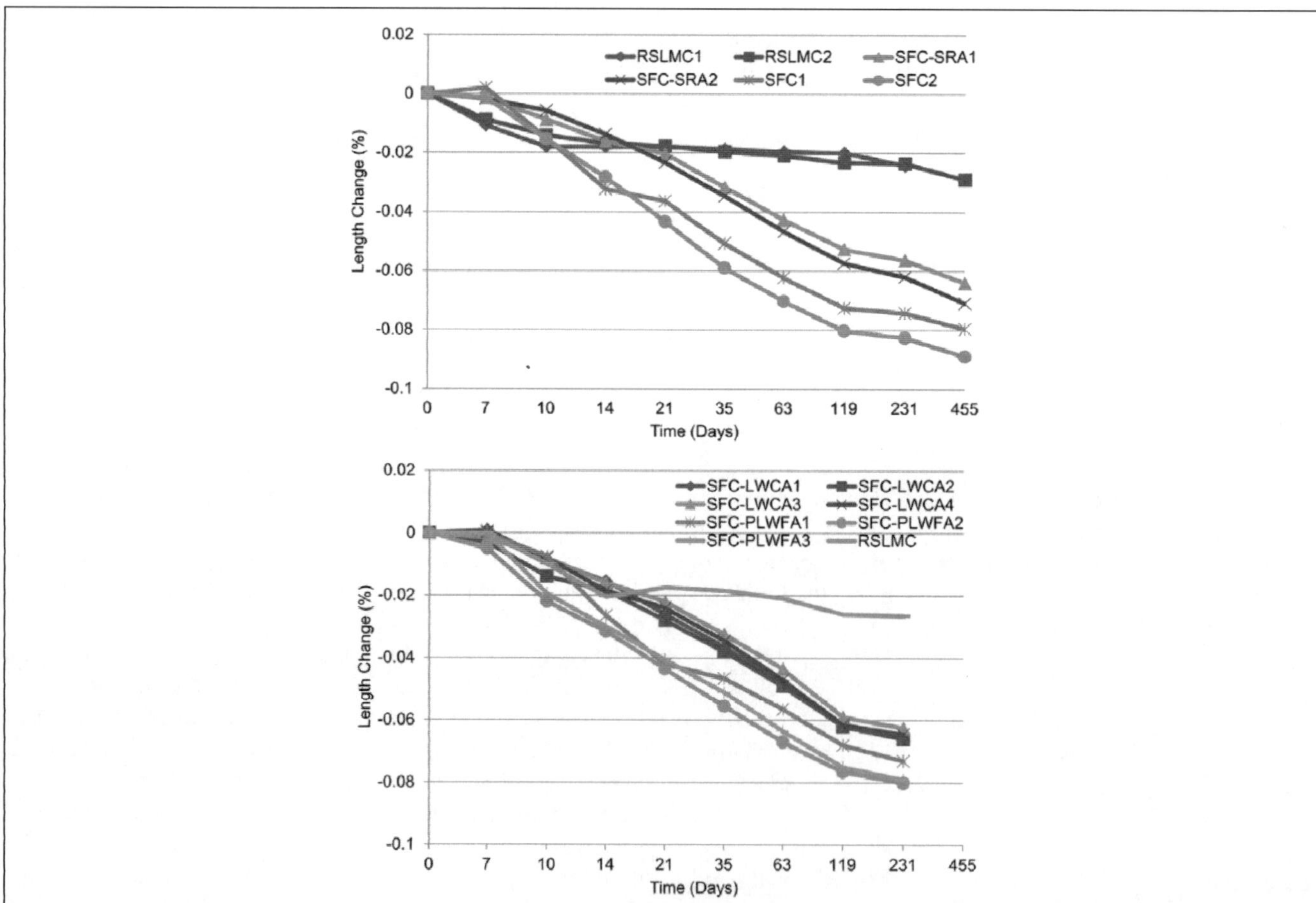

Figure 3. Shrinkage data for the westbound overlays (*top*) and the eastbound overlays (*bottom*): RSMLC = latex-modified concrete with rapid set cement; SFC = silica fume concrete; SFC–LWCA = SFC with lightweight coarse aggregate; SFC–PLWFA = SFC with partial lightweight fine aggregate; SFC–SRA = silica fume concrete with shrinkage reducing admixture.

Figure 4. Plastic shrinkage cracking in latex-modified concrete with rapid set cement in the left lane of the westbound bridge.

lane of the westbound bridge has extensive cracking, thought to be plastic shrinkage cracking (Figure 4). Even though good curing procedures were used, the weather conditions during placement were conducive to high evaporation rates, as shown with the high wind and low humidity in Table 3. This indicates the high sensitivity of RSLMC to curing conditions in adverse weather. In addition, a truck caught fire on that lane, possibly contributing to cracks.

Third Survey. A third condition survey was conducted on March 24, 2017. For each bridge condition survey, the traffic lane was closed to traffic. The cracks and delaminations in overlays in the traffic lanes were observed and measured. The crack survey results for overlays are summarized in Table 7. In general, there were few cracks except for the RSLMC in the westbound lane. The high cracking density for RSLMC, also observed in the previous survey, was attributed to weather conditions and the truck catching fire, as noted before. The cracks were narrow, most 0.1 mm or less, and a few were as wide as 0.2 mm. The cracks were in both the longitudinal and transverse directions. The passing lane was observed for cracking from the traffic lane; however, measurements were not taken.

The results indicated that the RSLMC and SFC overlays had spans with no cracking, but they also had other spans with cracks. This indicated the importance of curing in these overlays. There were also small spots repaired with epoxy mortar in the silica fume sections, as shown in Figure 5. These were caused by the mixing sequence leaving silica fume bags in the mixture that necessitated the removal of bag pieces and repairing of the holes left after removal.

Chain drag and hammer soundings indicated limited delaminated areas near the longitudinal joint, as indicated in Table 7. This was attributed to poor surface preparation leaving puddles of water that weakened the bond between the base concrete and the overlay. The hydrodemolition used had provided a rough and clean surface to bond; however, water puddles would adversely affect the bond.

Summary and Conclusion

- A RSLMC overlay is one of the best options to achieve very early strengths using special cement that provides a compressive strength of 3,000 psi within a few hours to limit lane closures and minimize inconvenience to the traveling public.
- The SFC overlays attained a compressive strength of about 3,000 psi, primarily in 3 days; some of the SFC overlays attained a strength of 3,000 psi in 1 day. The 3,000 psi gain in 1 day was in concretes with

Table 7. Cracks and Delaminations in the Overlays

	Cracks					Delaminations		
	Transverse		Longitudinal		Density (%)			
Type	No.	L (ft)	No.	L (ft)	L/Area	No.	Area (ft^2)	% Area
Westbound								
SFC–SRA	None	-	-	-	-	-	-	-
SFC–SRA	None	-	-	-	-	3	6	0.4
SFC	2	4	2	2	0.4	-	-	-
RSLMC	20	125.5	8	29.5	10.8	1	12	0.8
RSLMC	8	26.5	2	8	2.4	-	-	-
Eastbound								
SFC–LWCA	None	-	-	-	-	-	-	-
SFC–LWCA	7	11.5	-	-	0.8	-	-	
SFC–PLWFA	1	1	-	-	0.7	3	40.5	2.8
SFC–PLWFA	None	-	3	10.5	7.3	11	67	4.7
RSLMC	None	-	None	-	-	1	6	0.4

Note: L = length; SFC = silica fume concrete; SRA = shrinkage reducing admixture; LWCA = lightweight coarse aggregate; LWFA = lightweight fine aggregate; RSLMC = latex-modified concrete with rapid set cement; - = no data.

Figure 5. Small dark spots because of repair of holes left when bag pieces were removed.

lightweight aggregates and was attributed to a relatively low w/cm and internal curing.

- The SFC overlays with SRA, SFC with lightweight coarse aggregates, and SFC with lightweight fine aggregate are performing well with no or tight cracks after 2–3 years of exposure.
- The overlays had very tight cracks except for one section of the RSLMC in the left lane of the westbound bridge, which had extensive cracks attributed to plastic shrinkage from adverse weather conditions at placement. In addition, a truck had caught fire on that lane, which may have contributed to the development of cracks.
- SFC with SRA and lightweight coarse aggregates had low shrinkage values and exhibited tight cracks. SFC with no SRA and SFC with partial fine aggregate had shrinkage values above those recommended but still had tight cracks attributed to good curing practices and favorable weather conditions. Other factors effective in controlling cracking in these concretes are: (1) use of ternary cementitious materials in SFC with normal weight aggregates, (2) LW aggregate in SFC provides internal curing, and (3) all concretes had low elastic moduli.
- Dark spots were found in the SFC overlay in areas repaired after removal of the silica fume bag pieces. Adding silica fume in bags early in the loading sequence minimized this problem.
- Chain drag and hammer soundings indicated limited delaminated areas near the longitudinal joint, which were attributed to surface preparation leaving water puddles. Proper surface preparation practices are needed for overlay placement.

Acknowledgments

The authors thank the Virginia Transportation Research Council and staff of VDOT's Staunton District Structure & Bridge Section, especially Rex Pearce and Kim Ikhyeon; staff of VDOT's Central Office Structure & Bridge and Materials Divisions; and the FHWA. Particular thanks go to Park Thompson, Darrell Hayes, Evelina Khakimova, Michael Burton, Kenneth Herrick, William Ordel, Mary Bennett, Andy Mills, Deven Barkley, Mary Sharifiasl, Gail Moruza, and Liam Gallagher.

Author Contributions

The authors confirm contribution to the paper as follows: study conception and design: H. Celik Ozyildirim, Harikrishnan Nair; data collection: H. Celik Ozyildirim, Harikrishnan Nair; analysis and interpretation of results: H. Celik Ozyildirim, Harikrishnan Nair; draft manuscript preparation: H. Celik Ozyildirim, Harikrishnan Nair. Both authors reviewed the results and approved the final version of the manuscript.

References

1. Krauss, P. D., J. S. Lawler, and K. A. Steiner. *NCHRP Report 20-07/ Task 234: Guidelines for Selection of Bridge Deck Overlays, Sealers and Treatments*. Transportation Research Board of the National Academies, Washington, D.C., 2009.
2. Ozyildirim, C. Permeability Specifications for High-Performance Concrete Decks. *Transportation Research Record: Journal of the Transportation Research Board*, 1998. 1610: 1–5.
3. Sprinkel, M. Very-Early-Strength Latex-Modified Concrete Overlay. *Transportation Research Record: Journal of the Transportation Research Board*, 1999. 1668: 18–23.
4. Ozyildirim, C. Efforts to Improve Durability of Concrete in Virginia. *Transportation Research Record: Journal of the Transportation Research Board*, 2006. 1979: 79–83.
5. Sprinkel, M. Sampling and Testing Latex-Modified Concrete for Permeability to Chloride Ion. *Transportation Research Record: Journal of the Transportation Research Board*, 2009. 2113: 47–52.
6. Wang, K., D. C. Jansen, S. P. Shah, and A. F. Karr. Permeability Study of Cracked Concrete. *Cement and Concrete Research*, Vol. 27, No. 3, 1997, pp. 381–393.
7. Aldea, C. M., S. P. Shah, and A. Karr. Permeability of Cracked Concrete. *Materials and Structures*, Vol. 32, No. 5, 1999, pp. 370–376.
8. Lawler, J. S., D. Zampini, and S. P. Shah. Permeability of Cracked Hybrid Fiber Reinforced Mortar under Load. *ACI Materials Journal*, Vol. 99, No. 4, 2002, pp. 379–392.
9. Virginia Department of Transportation. *Road and Bridge Specifications*. Richmond, 2016.
10. Nair, H., C. Ozyildirim, and M. M. Sprinkel. *Use of Lightweight Concrete for Reducing Cracks in Bridge Decks*. Publication VTRC 16-R14. Virginia Transportation Research Council, Charlottesville, 2016.
11. Bentz, D. P., and W. J. Weiss. *Internal Curing: A 2010 State-of-the-Art Review*. Publication NISTIR Report 7765. National Institute of Standards and Technology, Gaithersburg, Md., 2011.
12. Ozyildirim, C. Durability of Structural Lightweight Concrete. Presented at ESCSI Special Workshop on Lightweight Aggregate Concrete Bridges, Concrete Bridge Conference, St. Louis, Missouri, 2008.

13. Sprinkel, M. M. *Overview of Latex Modified Concrete Overlays*. Report VHTRC 85-R1. Virginia Highway &Transportation Research Council, Charlottesville, 1984.
14. Ozyildirim, C. Laboratory Investigation of Concrete Containing Silica Fume for Use in Overlays. *ACI Materials Journal*, Vol. 84, No. 1, 1987, pp. 3–7.
15. Malhotra, V. M., V. S. Ramachandran, R. F. Feldman, and P. C. Aitcin. *Condensed Silica Fume in Concrete*. CRC Press, Boca Raton, Fla., 1987.
16.Babaei, K., and N. M. Hawkins. *STP 1100: Performance of Bridge Deck Concrete Overlays*. In *Extending the Life of Bridges*. ASTM, West Conshohocken, Pa., 1990.
17. Sprinkel, M. M. *High Early Strength Latex Modified Concrete Overlay*. Publication VTRC 88-R12. Virginia Transportation Research Council, Charlottesville, 1988.
18. Kuhlmann, L. A. Styrene-Butadiene Latex-Modified Concrete: The Ideal Concrete Repair Material? *Concrete International*, Vol. 12, No. 10, 1990, pp. 59–65.
19. Luther, M. D. Silica Fume (Microsilica) Concrete in Bridges. *Concrete International*, Vol. 15, No. 4, 1993, pp. 29–33.
20.Sprinkel, M. M. *Evaluation of Latex Modified and Silica Fume Concrete Overlays Placed on Six Bridges in Virginia*. Publication VTRC 01-R3. Virginia Transportation Research Council, Charlottesville, 2000.
21. Whiting, D. A., and R. Detwiler. *NCHRP Report 410: Silica Fume Concrete for Bridge Decks*. Transportation Research Board of the National Academies, Washington, D.C., 1998.
22Ramey, G. E., and R. S. Oliver. *Rapid Rehabilitation/ Replacement of Bridge Decks*. Highway Research Center, Harbert Engineering Center, Auburn University, Auburn, Ala., 1998.
23.Schemmel, J. J., J. C. Ray, and M. L. Kuss. *ACI SP-189: Impact of Shrinkage Reducing Admixture on Properties and Performance of Bridge Deck Concrete*. American Concrete Institute, Farmington Hills, Mich., 2000.
24. Maggenti, R., C. Knapp, and S. Fereira. Controlling Shrinkage Cracking. *Concrete International*, Vol. 35, No. 7, 2013, pp. 36–41.
25. Babaei, K., and A. Fouladgar. Solutions to Concrete Bridge Deck Cracking. *Concrete International*, Vol. 19, No. 7, 1997, pp. 34–37.

The Standing Committee on Durability of Concrete (AFN30) peer-reviewed this paper (18-04566).

TRR
JOURNAL OF THE TRANSPORTATION RESEARCH BOARD

Article

Transportation Research Record
2018, Vol. 2672(27) 88–98

DOI: 10.1177/0361198118773198
journals.sagepub.com/home/trr
SAGE

Evaluation of Selected Durability Properties of Portland Cement Concretes Containing Ground Glass Fiber as a Pozzolan

Hassan Rashidian-Dezfouli[1] and Prasada Rao Rangaraju[1]

Abstract

Glass fibers are produced in large quantities around the world for the production of fiber glass composites and for other reinforcement applications. However, considering the stringent quality requirements a significant quantity of waste is produced in the production process that does not meet required specifications of the final product in terms of the physical properties, although the chemical composition of the glass fiber waste is identical to that of the in-spec glass fiber. At present, this waste is largely disposed of in landfills. This glass fiber waste has the same chemical composition as the good glass fiber, and is rich in silica, alumina, and calcium oxides. Considering the requirements of a typical pozzolanic material, waste glass fiber could be ground to a fine powder for use as a pozzolan. In this study, the effect of using ground glass fiber (GGF) as a pozzolan in concrete at different cement replacement levels (10, 20, and 30% by mass) on selected durability properties of mixtures over a span of two years was investigated. For this purpose, the resistance of GGF-containing mixtures against the alkali–silica reaction, sulfate attack, and drying shrinkage was evaluated and compared with two control mixtures, including a mixture with 100% Portland cement and a mixture having 25% class F fly ash. Results of this research showed that the replacement of Portland cement with GGF resulted in significant improvements in the durability properties, and the mechanical properties were comparable with those of control mixtures.

The durability of Portland cement concrete against aggressive environments is one of the most important factors affecting its long-term performance. Utilization of supplementary cementitious materials (SCMs) such as fly ash, slag, and silica fume in Portland cement mortar and concrete mixtures has been widely studied and reported to be beneficial in improving their durability properties (*1–5*).

Ground glass fiber (GGF) is a fine powder that is produced by finely grinding waste fiber glass. Like other SCMs, GGF has a high amount of silica, alumina, and iron oxides (i.e., $S + A + F > 50\%$) and can be considered as a pozzolan. Unlike ground glass powder, which is produced from milling recycled soda-lime glass and has a high alkali content, the alkali content of GGF is very low. According to earlier studies, because of the high alkali content of the material, the susceptibility of the ground glass powder-containing concrete mixtures to the alkali–silica reaction (ASR) has always been a big concern (*6–9*). However, because of its low alkali content, GGF could potentially be used as a pozzolan without concerns in increasing ASR damage.

Utilization of waste generated from the glass fiber production as an SCM in concrete mixtures is a relatively new concept. The few studies that have been conducted on this subject have shown the advantages of using this material in improving fresh and hardened properties of mortar and concrete mixtures. Chen et al. investigated the utilization of E-glass particles in Portland cement mixtures (*10*). It was reported that replacing cement with ground E-glass, to 40% by mass of the cement, led to an increase in the strength activity index (SAI) of mixtures. In addition, use of this material improved durability properties of concrete such as resistance against ASR, sulfate attack, and chloride ion penetration. In recent studies, Rangaraju et al. and Rashidian-Dezfouli and Rangaraju investigated the utilization of GGF as an SCM in concrete and mortar mixtures (*11, 12*). It was observed that

[1]Glenn Department of Civil Engineering, Clemson University, Clemson, SC

Corresponding Author:
Address correspondence to Prasada Rao Rangaraju: prangar@clemson.edu

Table 1. Chemical Composition and Physical Properties of the Cementitious Materials

Chemical composition/physical properties	Cement (%)	GGF (%)	Fly ash (%)
Physical properties			
Specific gravity	3.15	2.60	2.25
Blaine's fineness (cm^2/g)	4,720	10,200	6,040
Loss on ignition (LOI)	2.6%	1.0%	2.3%
Chemical composition			
SiO_2	19.9	47.7	50.7
CaO	62.3	19.6	3.3
Al_2O_3	4.8	10.4	25.1
Fe_2O_3	3.1	0.3	12.5
MgO	2.7	2.3	1.1
$Na_2O_{eq.}$	0.38	0.70	2.0
SO_3	3.23	0.02	0.70

Figure 1. (*a*) Appearance of GGF and (*b*) the SEM image of GGF.

replacing cement with GGF to 30% (mass-based) did not affect the mechanical properties of specimens. However, there was a significant improvement in short-term durability properties such as resistance against ASR (*11, 12*), sulfate attack, and chloride ion penetration (*12*).

This paper presents findings from selected durability studies that were conducted on GGF-containing (10, 20, and 30% of the cement mass) concrete and mortar specimens for two years. For comparison purposes, results were compared with two control mixtures, including a mixture with 100% Portland cement, and a mixture having 75% Portland cement and 25% class F fly ash. Tests conducted in this study include the resistance of concrete bar against ASR (ASTM C1293), resistance against sulfate attack (ASTM C1012), and drying shrinkage (ASTM C596). Furthermore, to study the microstructure of specimens as well as the mechanisms involved, thermogravimetric analysis (TGA), scanning electron microscopy (SEM), and X-ray diffraction analysis (XRD) were performed. Whereas earlier studies have shown the good mechanical properties of GGF-containing specimens (*11, 12*), the results of this study could be beneficial in evaluating the long-term durability properties of GGF-containing concrete and mortar specimens.

Experimental Program

Materials

For this study the following materials were used.

Cement. A Type I/II (ASTM C 150) ordinary Portland cement with a specific gravity of 3.15 and an average particle size of 17 μm was used in this study. Selected physical properties and the chemical composition of the Portland cement are presented in Table 1. The bogue composition of the cement is C_3S = 65.7%, C_2S = 7.5%, C_3A = 7.5%, and C_4AF = 9.4%.

Ground Glass Fiber. The GGF used in this study was a fine white powder that was prepared by milling off-spec glass fiber in a ball mill to a fine powder. This material has the specific gravity of 2.60 and an average particle size of 4 microns (Figure 1*a*). The SEM observation of GGF particles is presented in Figure 1*b*. As can be seen, GGF particles are angular and have a smooth surface texture. Selected physical properties and the chemical composition are presented in Table 1.

Fly Ash. For comparison purposes, a class F fly ash (ASTM C618) with a specific gravity of 2.25 and an average particle size of 28 microns was obtained from a local source and used as an SCM. Selected physical properties and the chemical composition are presented in Table 1.

Fine Aggregate (Siliceous Sand). A non-reactive siliceous river sand with a fineness modulus of 2.60 and an absorption value of 0.30% with an oven-dry specific gravity of 2.67 was used in this study.

Coarse Aggregate. In order to investigate the resistance of concrete mixture against the ASR, a highly reactive rhyolitic aggregate from Las Placitas gravel pit from New Mexico (NM) was used. The water absorption and the oven-dry specific gravity of this aggregate is 1.10% and 2.60, respectively.

Mixture Properties

In this study, GGF was employed as an SCM at different replacement dosage levels of 10, 20, and 30% by the mass of the cement. In addition, two control mixtures, including a mixture consisting of 100% plain Type I Portland cement as the binder (CTRL-1) and a binary blend of 75% Type I/II Portland cement with 25% Class F fly ash, by mass, as the binder (CTRL-2), were tested for comparative purposes. For the tested mortar and concrete mixtures, the mix ID was selected based on the composition of the cementitious portion of each mixture as presented in Table 2. To select the water-to-binder ratio and the aggregate content, the relevant test method was followed to proportion the mixture.

Experimental Test Methods

Pozzolanic Activity. To evaluate the pozzolanic reactivity of GGF, SAI and TGA was conducted on mortar cubes and samples of pastes, respectively.

SAI. Following the ASTM C311, the pozzolanic reactivity of GGF was investigated by measuring the SAI of GGF-containing mortar cubes. In this test, 50 × 50 × 50 mm (2 × 2 × 2 in.) mortar cubes were cast for GGF-10, GGF-20, and GGF-30 mixtures and the results were compared with the SAI for CTR-1 and CTRL-2 mixtures after 7 and 28 days.

TGA. The efficiency of GGF in reducing the calcium hydroxide (CH) of the paste through the pozzolanic reaction was investigated by performing TGA on CTRL-1 and GGF-containing paste samples. To perform this test, each paste sample was crushed into a fine powder and was then heated from ambient temperature to 700°C (1292°F). The weight change of the sample was monitored and recorded during the test. It is known that the decomposition of CH takes place in the temperature range of 400°C to 520°C (752°F to 968°F), which causes a significant mass loss in the sample. The mass loss that happens in this temperature range can be used to calculate the amount of CH of the sample. In this study, TGA was performed on 28 and 56-day paste samples.

Absorption Test. In this study, an absorption test was performed on two-year-old mortar bars from all the mixtures. For this test, a similar test procedure as presented in ASTM C1403 was followed. Before the test, mortar specimens were dried in an oven at 110°C (230°F) for 24 h to remove all their moisture. Samples were then removed from the oven, and the initial mass of each sample was recorded. Subsequently, mortar samples were placed in sealed containers in a manner that they were sitting on two supports so that 3 mm (0.12 in.) of samples would penetrate into water. Containers were sealed and left at ambient temperature for 24 h. After this period, specimens were removed from the container, and their surface dry mass was measured and recorded. The change in the weight of mortar samples can be used to indicate the absorption of each mixture. It should be noted that although ASTM C1403 states the use of mortar cubes, in this study, the absorption test was conducted on mortar bars that were used for the drying shrinkage experiment.

Resistance against ASR. The effectiveness of GGF in mitigation of ASR expansion was evaluated by performing

Table 2. Mixture ID and the Composition of Binder Portion of Pastes, Concretes and Mortars

Mix ID	CTRL-1	CTRL-2	GGF-10	GGF-20	GGF-30
Cement	100%	75%	90%	80%	70%
Fly ash	–	25%	–	–	–
GGF	–	–	10%	20%	30%

the concrete prism test (CPT, ASTM C1293). For this test, NM aggregate was used as the reactive coarse aggregate and a non-reactive sand was used as the fine aggregate. Following ASTM C1293, mixtures were cast in 75 × 75 × 285 mm (3 × 3 × 11.25 in.) molds, and then cured for 24 h in a moist room at 23°C (73°F) and 100% relative humidity (RH). Specimens then were demolded and the initial length measurement was performed on each specimen. Finally, concrete prisms were placed in sealed containers and rested in a 38°C (100°F) chamber. To monitor the expansion of the concrete prisms, comparator readings were performed on each prism at the time span specified by the ASTM C1293 up to two years.

Sulfate Attack Resistance. Following ASTM C1012, 25 × 25 × 285 mm (1 × 1 × 11.25 in.) mortar bars and sufficient numbers of 50 × 50 × 50 mm (2 × 2 × 2 in.) mortar cubes were cast for each mixture. Specimens were removed from their molds after 24 h and placed in a saturated lime water bath until mortar cube specimens reached the required compressive strength of 20.7 MPa (3000 psi). At this time, the length measurement of specimens was started by taking the zero-day reading. Specimens were then placed in a 5% sodium sulfate solution and the expansion of mortar bars was monitored up to two years. The old solution was replaced with a fresh one after each reading.

Resistance against Drying Shrinkage. Drying shrinkage of mortars bars was measured in accordance with ASTM C596. For this experiment, three mortar bars from each mixture were cast in 25 × 25 × 285 mm (1 × 1 × 11.25 in.) prism molds. Specimens were kept in a moisture room for 24 h and then were demolded and placed in a saturated lime solution for an additional 48 h. After this period, mortar bars were removed from the solution, their lengths were measured, and then were placed in the environment chamber until the next reading. The temperature and RH of the chamber were kept at 23°C (73°F) and 50%, respectively. In order to obtain the drying shrinkage percent, length change of mortar bars was measured periodically up to 24 months.

Summary of the Test Plan

A summary of the test plan, related standard, number of specimens that were cast for each mix ID, and the water-to-binder ratio used for each experiment are presented in Table 3. All of the tests were conducted in duplicate, and the results from the tests were found to be within accepted limits of variability in standard test methods.

Results and Discussion

Pozzolanic Reactivity

SAI. Table 4 presents results of the compressive strength of mortar cubes and the resultant SAI values after 7 and 28 days. As can be observed, the SAI value for the GGF-20 mixture is 104% and 110% after 7 and 28 days, respectively; both are higher than 75%, and so GGF can be considered as a pozzolan, based on ASTM C618. Furthermore, for all GGF-containing mixtures, SAI

Table 3. Summary of the Test Plan

Name of the experiment	Standard used	Number of specimens (for each Mix ID)	Water-to-binder ratio				
			CTRL-1	CTRL-2	GGF-10	GGF-20	GGF-30
Strength activity index (SAI)	ASTM C311	3 (for each age)	0.48	0.45	0.48	0.47	0.46
Resistance against ASR	ASTM C1293	3	0.45	0.45	0.45	0.45	0.45
Resistance against sulfate attack	ASTM C1012	3	0.49	0.45	0.48	0.47	0.46
Drying shrinkage	ASTM C596	4	0.44	0.40	0.42	0.40	0.39

Table 4. Compressive Strength and Strength Activity Index of Mortar Cubes after 7 and 28 Days (1 MPa is equal to 145 psi)

Mix ID	7 days		28 days	
	Compressive strength (MPa)	SAI (%)	Compressive strength (MPa)	SAI (%)
CTRL-1	42	100	54	100
CTRL-2	34	81	48	89
GGF-10	38	89	56	103
GGF-20	44	104	60	110
GGF-30	41	97	58	106

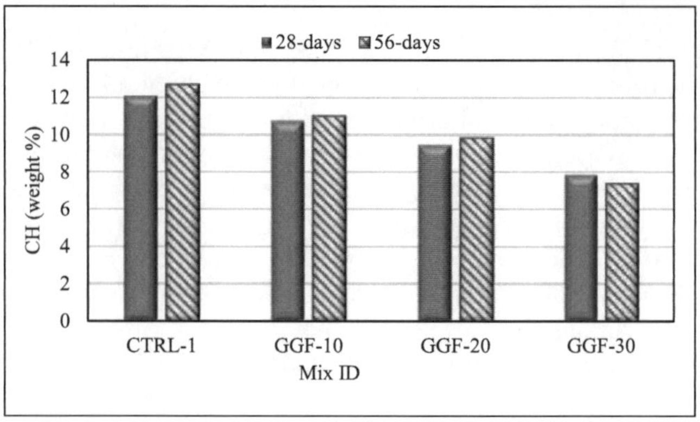

Figure 2. Effect of GGF dosage on the CH content of paste samples.

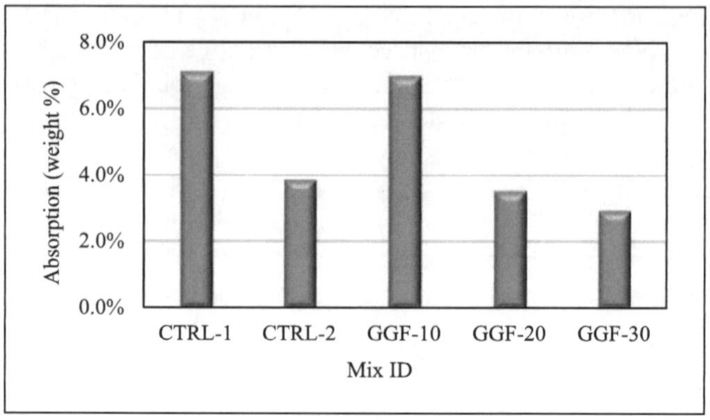

Figure 3. Water absorption of mortar specimens.

values of 7 and 28-day specimens were higher than the CTRL-2 mixture, which has 25% fly ash.

The better performance of GGF-containing mixtures compared with the CTRL-2 mixture may be related to the smaller particles size of GGF (4 μm in comparison to fly ash (28 μm particles. Because of its smaller particle size and higher specific surface, GGF can refine the porosity of the binder's matrix and can also engage in the pozzolanic reaction faster. Furthermore, XRD results showed that GGF has a very amorphous structure, whereas the presence of crystalline phases was obvious in fly ash. This amorphous structure of GGF promotes the pozzolanic reaction by allowing silica to react with CH to form calcium silicate hydrate (C-S-H) gel. As a result of this reaction, and the refinement effect of GGF particles, SAI values of all GGF-containing mixtures surpassed the SAI of the CTRL-1 mixture after 28 days (see Table 4).

TGA Results. The CH content of 28-day and 56-day-old CTRL-1 and GGF-containing paste samples are presented in Figure 2. As can be seen, the amount of CH content decreased significantly with the increase in GGF dosage level. Whereas the CH content of CTRL-1 paste sample was 12.1% and 12.7%, the CH content of GGF-30 sample was measured to be 7.9% and 7.4% after 28 and 56 days of curing, respectively. The reduction in the CH content could be caused by the dilution effect of GGF, as well as the pozzolanic reaction that consumes CH and generates C-S-H gel. Despite the overall reduction in the CH content, it should be noted that because of the small particle size of GGF, it could increase the cement hydration by providing more reaction sites, which increases the CH production rate at early stages. An increase in the normalized CH content (normalized for CH weight divided by per gram of cement) was reported in 28-day paste samples when a powder produced from glass fiber wastes (with the average size of 8 μm) was used as pozzolan (*13*).

Absorption Test

In order to investigate the effect of GGF in refining the porosity of mortar specimens, an absorption test was performed on mortar bars in accordance with the method presented earlier. The result of this test is presented in Figure 3. As can be seen, the addition of GGF caused a significant reduction in the water absorption of samples. Whereas the absorption of CTRL-1 mixture was 7.2%, the absorption of GGF-30 mixture was 2.9%. In addition, as is obvious from this figure, replacement of cement with fly ash also reduced the absorption value of the mixture significantly; however, the reduction amount was not as high as the GGF-20 mixture, which shows the higher efficiency of GGF in reducing the porosity of the specimens.

ASR

In this study, the efficiency of GGF in the mitigation of the ASR expansion of concrete specimens was evaluated by conducting the CPT (ASTM C1293). Expansion results of the CPT are presented in Figure 4. As can be seen, CTRL-1 samples showed a rapid expansion in the early stages and reached 0.08% expansion after 90 days. The expansion continued, at a lower rate, and reached 0.14% after two years. Although the addition of GGF caused a significant reduction in the expansion level, the GGF-10 mixture was not able to mitigate the expansion and showed expansion of 0.08% after two years. Further increase in the GGF dosage resulted in the reduction of the ASR-related expansion, and both GGF-20 and GGF-30 showed an expansion level of 0.02% after two years. In addition, CTLR-2 mixture was also able to mitigate the ASR expansion and showed an expansion level of 0.03. Figure 5 shows an SEM image of a polished section of CTRL-1 concrete prism that was exposed to the CPT condition for two years. As can be seen, the aggregate is cracked as a result of the formation of ASR gel.

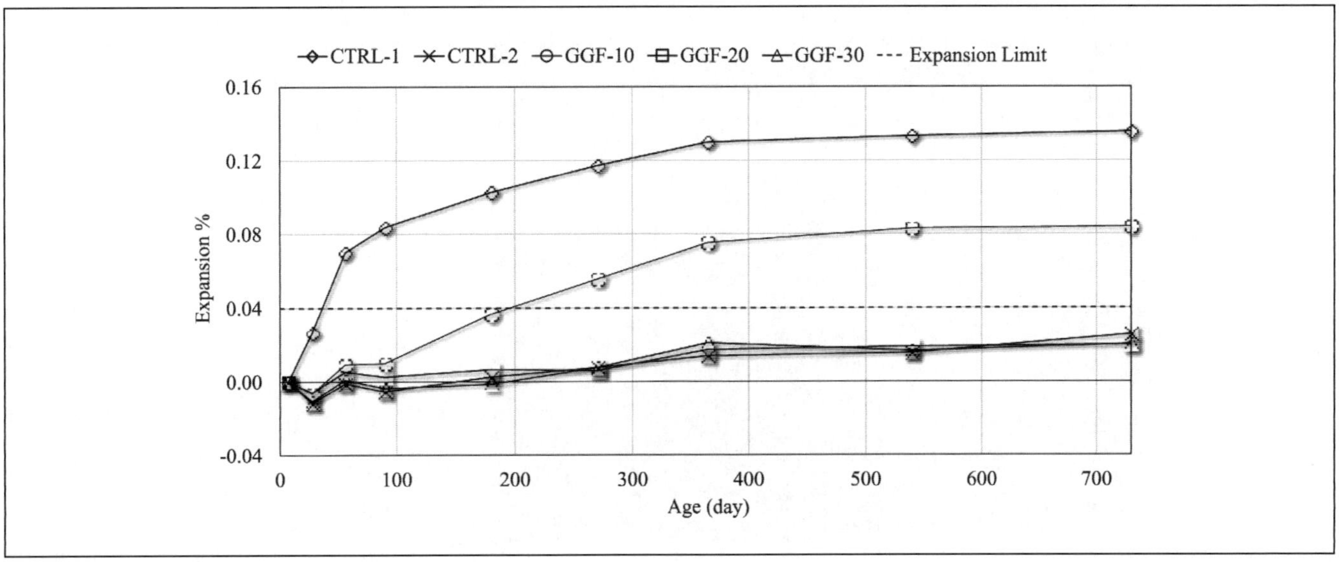

Figure 4. Expansion of concrete prism versus time.

The main mechanisms for the mitigation can be summarized as follows.

Reduction in Available Calcium for Newly Formed ASR Gel. As was shown in Figure 2, the use of GGF as an SCM in Portland cement mixtures reduces the CH level of the paste. It is known that the presence of CH is essential for the calcification of the fresh ASR gel. Calcification of the ASR gel increases its viscosity, and the resultant ASR gel can induce significant tensile stress on the mortar matrix on absorption of moisture and expansion (*3, 14*). As a result, the addition of GGF or fly ash will reduce the risk of ASR damage by lowering the CH level of the paste, thus reducing the ASR expansion and damage.

Reduction in Ca/Si Ratio. Utilization of GGF and fly ash in Portland cement systems increases the amount of amorphous silica content in these mixtures. This leads to the formation of a C-S-H gel with a lower Ca/Si ratio as a result of the pozzolanic reaction or refining the structure of the existing C-S-H gel. According to earlier studies, C-S-H gels with lower Ca/Si ratio will retain a higher amount of alkalis in comparison with the gels with greater Ca/Si ratio (*2, 15*). Therefore, it is likely that the C-S-H gel in the GGF or fly ash-containing samples retains more alkalis in its structure.

The effect of GGF or fly ash as SCMs in lowering the Ca/Si ratio of the paste matrix was investigated by performing energy-dispersive X-ray spectroscopy (EDX) on CTRL-1, CTRL-2, and GGF-30 samples. For this purpose, the weight ratio of Ca and Si from at least seven spectrums from different locations of each sample was measured. The averaged Ca/Si ratio of these samples was calculated as 2.56, 2.27, and 1.76 for CTRL-1, CTRL-2, and GGF-30, respectively.

Effect of Alumina in Reducing ASR. Earlier studies have shown that the presence of Al, coming from SCMs, can improve the resistance of SCMs containing Portland cement mixtures against ASR expansion (*16, 17*). Two main mechanisms by which Al plays its role in reducing the ASR are: (i) the superior performance of calcium alumino-silicate hydrate (C-A-S-H) gel in binding alkalis of the pore solution in comparison with C-S-H gel (*18*); and (ii) reduction in the Si dissolution rate from the reactive aggregate into the pore solution (*19*). Regardless of the exact mechanism, an increase in the Al level by the addition of Al-rich SCMs can be beneficial in reducing ASR expansion of Portland cement systems. Therefore the better performance of GGF and fly-containing samples can be explained by their pastes' higher Al content. The EDX results performed on at least seven spectrums from different locations of each sample showed the average weight percent value of Al was 4.56%, 2.83%, and 4.56% for the case of CTRL-1, CTRL-2, and GGF-30 specimens, respectively.

Reduction in the Permeability of Samples. The GGF used in this study has a very fine particle with the average size of 4 μm. Addition of this material, especially at higher replacement levels such as 20% and 30%, can refine the structure of the paste, reducing the porosity of samples. In addition, both fly ash and GGF introduce a considerable amount of amorphous silica into the mixtures, which can promote the pozzolanic reaction, causing further refinement of the paste structure. Refinement of the paste structure reduces the permeability of the sample. As presented in Figure 3, the absorption of mortar specimens decreased with the addition of fly ash or GGF. Increase in the weight of the mortar specimens in the absorption

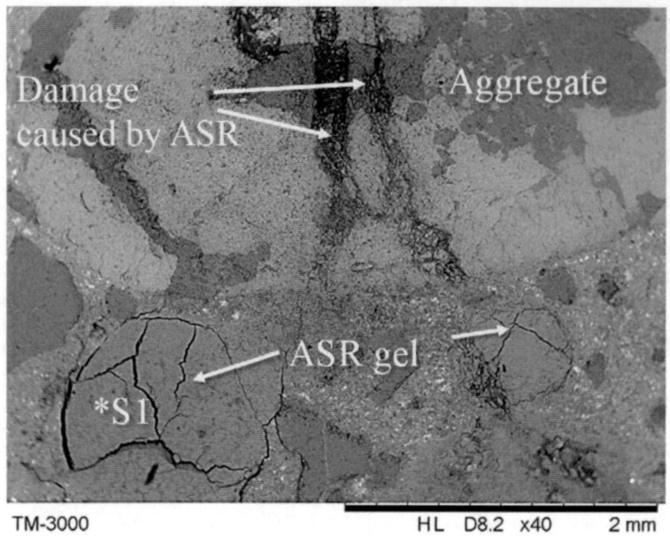

Element	S1 (weight %)
Na	9.79
Mg	0.31
Al	1.06
Si	56.41
K	12.43
Ca	18.87
Fe	1.09

Figure 5. ASR damage in CTRL-1 sample.

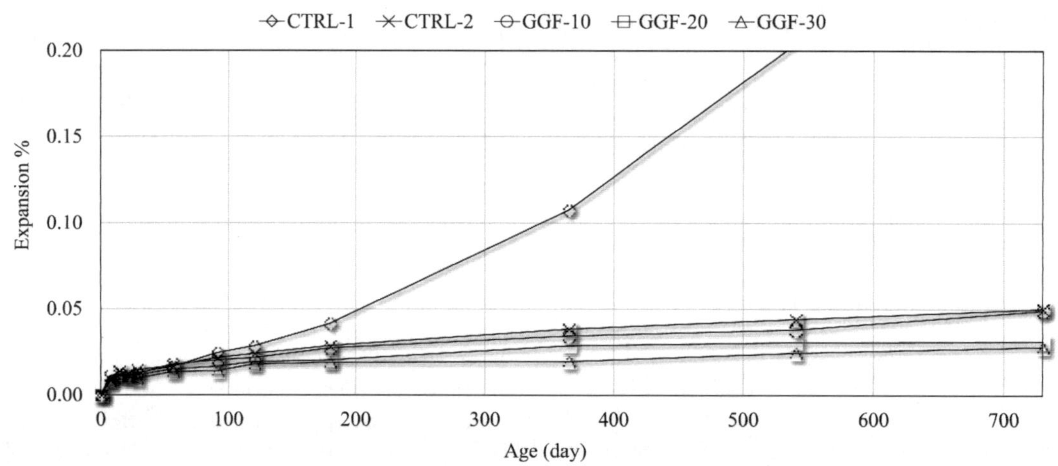

Figure 6. Expansion results of mortar bars subjected to 5% sodium sulfate solution.

test was 7.1%, 3.9%, and 2.9% for the case of CTRL-1, CTRL-2, and GGF-30, respectively.

Sulfate Attack

Expansion results of mortar bars that were exposed to a 5% sodium sulfate solution for two years are presented in Figure 6. As can be seen in this figure, the utilization of GGF in an SCM effectively reduced the expansion of mortar bars. After two years, the expansion of CTRL-1 sample reached 0.27%, whereas the expansion value of CTRL-2, GGF-10, GGF-20, and GGF-30 was 0.05%, 0.05%, 0.03%, and 0.03%, respectively. The main mechanisms associated with the reduction in the expansion of SCM-containing mortar bars could be related to: (i) the lower amount of CH in the paste structure (Figure 2), which reduces the risk of the formation of gypsum and ettringite that could cause the expansion; and (ii) the lower absorption of this sample as a result of the refinement of the paste structure, which could hinder the penetration of the sulfate solution into the matrix.

The expansion of Portland cement-based specimens exposed to the sodium sulfate solution has been related to the formation of gypsum and ettringite that are generated by the reaction of sulfur ions of the solution and CH of the paste (*20, 21*). To examine the formation of gypsum and ettringite in the mortar sample, an XRD test was performed on CTRL-1, CTRL-2, and GGF-30 specimens after being exposed to the sodium sulfate solution for one year. XRD patterns of these samples are presented in Figure 7. In this figure, the formation gypsum and ettringite is evident from the XRD pattern of the CTRL-1 samples. However, the peaks associated with gypsum and ettringite showed a very low intensity, indicating the lower formation these in CTRL-2 and GGF-30 samples in comparison to CTRL-1.

To investigate the penetration of sulfur ions into the exposed samples, EDX analyses were performed on at least five points at each selected location of the samples. The location was selected to cover the different depth of the samples, from 0 μm (at the edge of the sample) to 1000, 2000, and 3000 μm from the edge of the sample toward the center of the sample. The results for the weight percent of the sulfur are summarized in Figure 8. As can be seen in this figure, in comparison with CTRL-1, CTRL-2 and GGF-30 samples showed a lower weight percent of sulfate ions, particularly at more inner depths (i.e., from 2000 to 3000 μm).

Drying Shrinkage

Results of the drying shrinkage test are presented Figure 9. As shown in this figure, despite the increase in the drying shrinkage values of GGF-containing mortar bars, overall they showed similar shrinkage values to the CTRL-1 sample. The higher drying shrinkage in GGF-containing samples in comparison with the CTRL-1 mixture can be explained by factors such as the higher paste volume of GGF-containing mixtures (lower specific gravity of GGF in comparison with Portland cement), and the paste refining effect of GGF, leading to a lower pore size when compared with CTRL-1 sample containing 100% cement. The lower pore size creates a higher shrinkage force when the sample is losing water, resulting in greater shrinkage in GGF-containing mortar bars. On the other hand, the low shrinkage of fly ash-containing mixture (CTRL-2) can be explained by its lower water demand (as presented in Table 3). The ASTM C596 test method requires the water-to-binder ratio of mixtures to be adjusted to achieve the value of 110 ± 5%. Thus, the smaller drying shrinkage of CTRL-2 in comparison with CTRL-1 can be explained by its lower water-to-binder ratio.

Conclusion

Based on the findings from the limited scope of this study, effects of utilization of GGF as an SCM in Portland cement concrete can be summarized as follows:

1. Based on the SAI and TGA results, GGF can be considered as a pozzolan. For all replacement levels (i.e., 10, 20, and 30%), SAI values of GGF-containing specimens were higher than 100% at 28 days. In addition, TGA results corroborate the SAI values by showing the reduction in the CH content of paste samples containing GGF pozzolan.
2. Increase in the dosage level of GGF significantly reduced the moisture absorption of the mortar specimens. According to the absorption results, whereas CTRL-1 specimens showed a weight gain of 7.2%, GGF-20 and GGF-30 showed a weight gain of only 3.5% and 2.9%, respectively. The decrease in the absorption of these samples could be explained by the small particle size of GGF that refines the microstructure of the paste.
3. According to the CPT results, at dosages of 20% and 30% GGF was able to mitigate ASR-induced expansion in concrete prisms. After two years, CTRL-1 and CTRL-2 samples showed an expansion of 0.140% and 0.026%, respectively, whereas GGF-20 and GGF-30 samples showed an expansion of only 0.020%, which was below the expansion limit of 0.040%. The lower expansion of GGF-containing samples can be explained by the lower CH content, lower Ca/Si ratio, higher Al content in the paste, and lower absorption of these samples in comparison with CTRL-1 specimens. The beneficial performance of GGF at 20% and 30% dosage levels in CPT was also found to be in line with the findings from AMBT studies reported by Rashidian and Rangaraju (*11*).
4. Addition of GGF significantly improved the resistance of mortar bars against sulfate attack. While CTRL-1 samples showed an expansion of 0.27%, expansion values of GGF-10, GGF-20, and GGF-30 were 0.05%, 0.03%, and 0.03%, respectively. The lower expansion values in GGF-containing samples could be related to the refined microstructure of the paste in these samples that hinders the penetration of sulfates into the matrix.
5. Drying shrinkage of mortar bars was found to increase with the increase in the replacement levels of GGF. This could be caused by the increased volume of the paste as a result of lower specific gravity of GGF as well as finer pore size in GGF-containing samples in comparison with CTRL-1.

Based on the findings from this study, the use of GGF as an SCM is a viable alternative to use of fly ash in concrete without losing any advantages that fly ash typically offers to concrete.

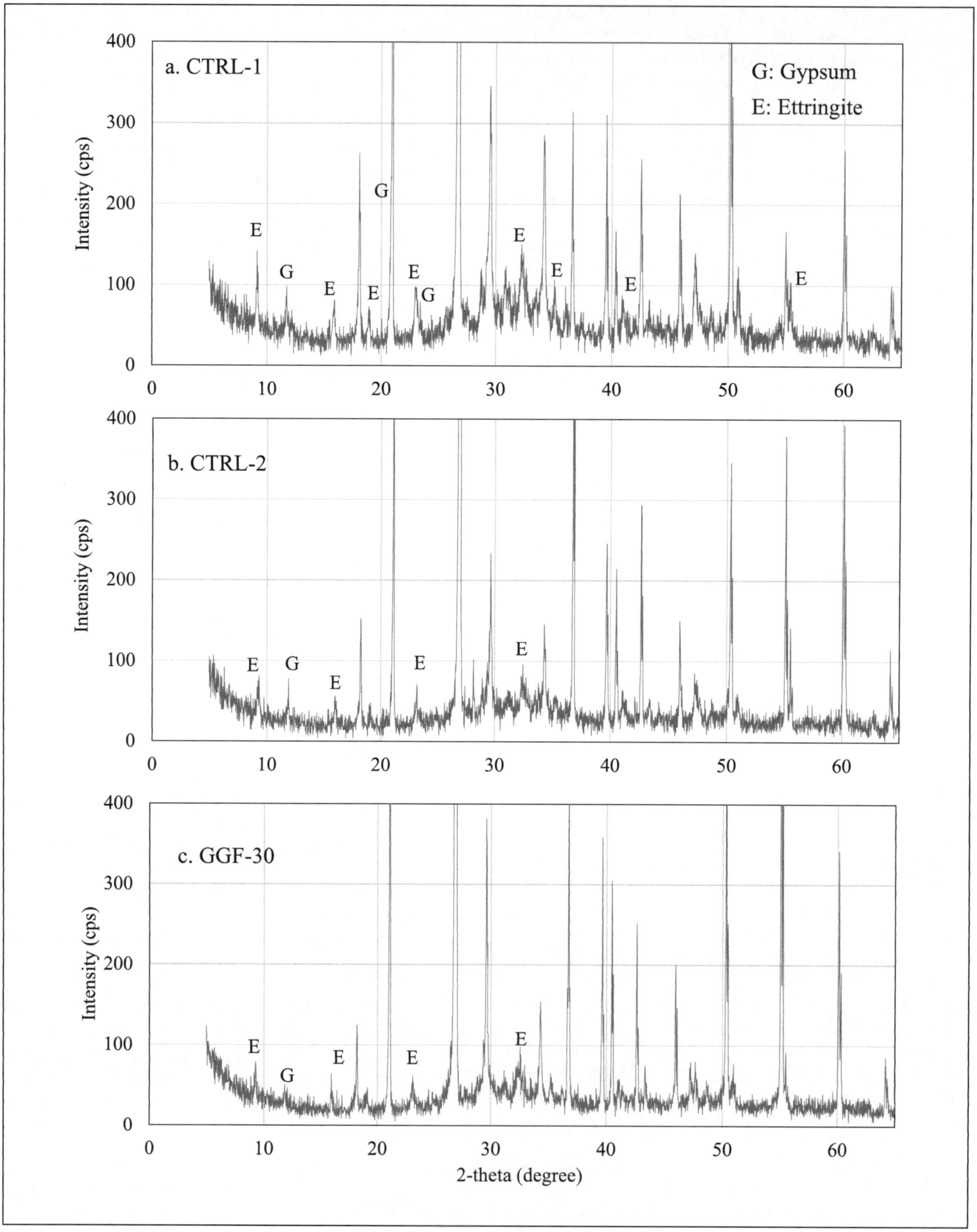

Figure 7. XRD pattern of samples exposed to 5% sodium sulfate solution: (a) CTRL-1; (b) CTRL-2; (c) GGF-30.

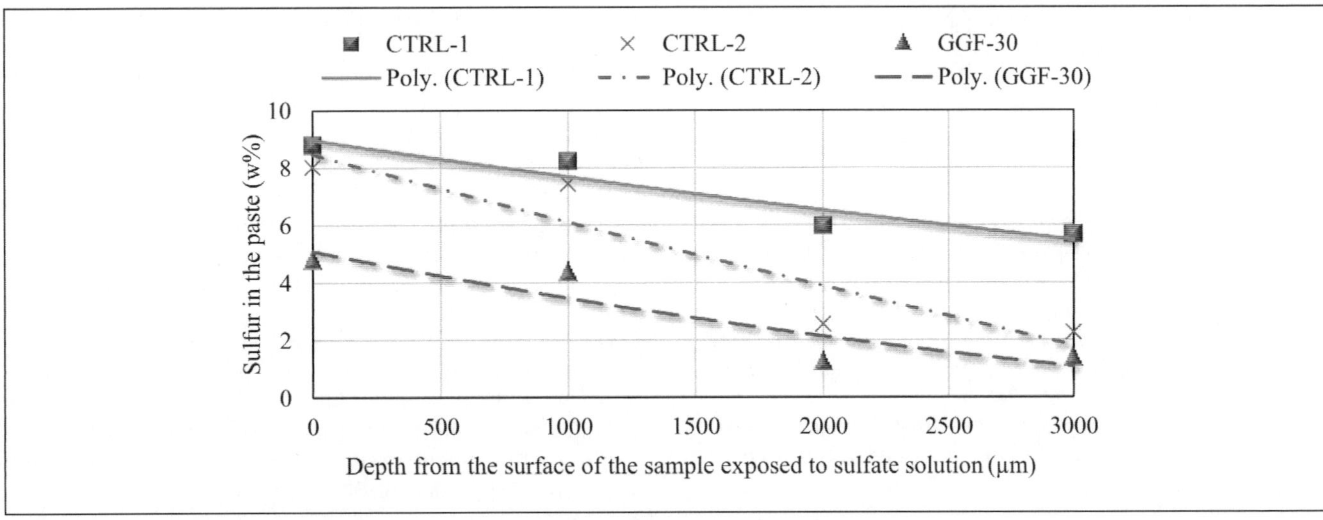

Figure 8. Concentration of sulfur ions in mortar bar samples (weight %).

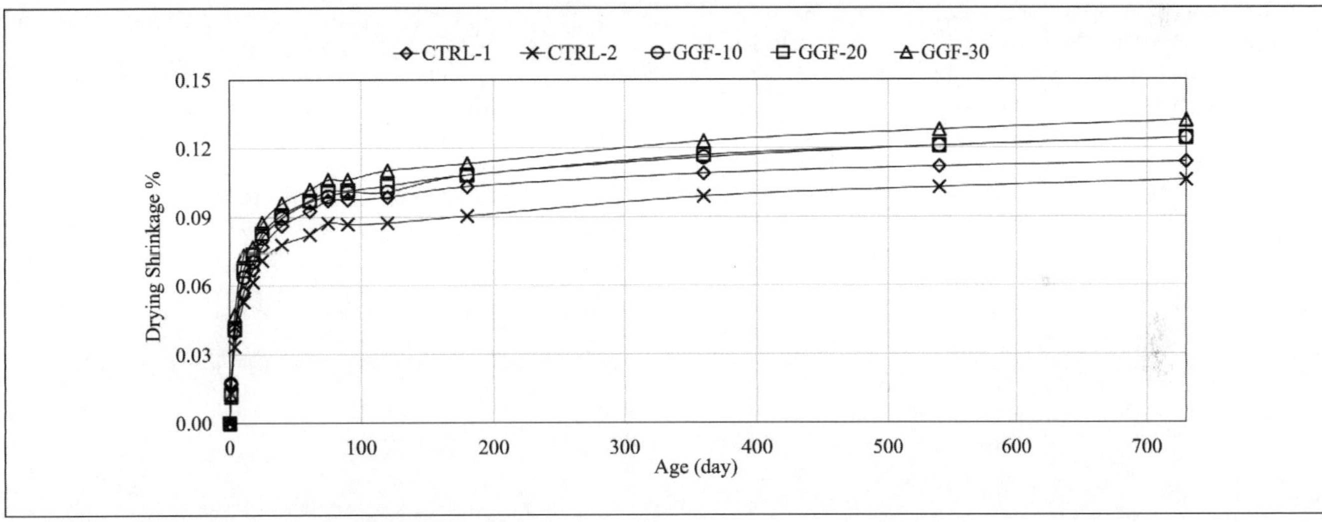

Figure 9. Shrinkage of mortar bars versus time.

Acknowledgments

This study was possible through funding from PPG, Inc. and ARGOS USA Corporation. In particular, thanks are extended to Dr. Gordon Nameni of PPG, Inc. and Mr. Godwin Amekuedi of ARGOS USA Corporation for their thoughtful comments.

Author Contributions

The authors confirm contribution to the paper as follows. Study conception and design: Rangaraju and Rashidian-Dezfouli; data collection: Rashidian-Dezfouli; analysis and interpretation of results: Rashidian-Dezfouli and Rangaraju; draft manuscript preparation: Rashidian-Dezfouli and Rangaraju. All authors reviewed the results and approved the final version of the manuscript.

References

1. Thomas, M. D. A., B. Fournier, and K. J. Folliard. *Alkali-Aggregate Reactivity (AAR) Facts Book*. FHWA-HIF-13-019. Federal Highway Administration, 2013.
2. Bhatty, M. S. Mechanism of Pozzolanic Reactions and Control of Alkali-Aggregate Expansion. *Cement, Concrete and Aggregates*, Vol. 7, No. 2, 1985, pp. 69–77.
3. Detwiler, R. The role of fly ash composition in reducing alkali-silica reaction. Portland Cement Association R&D Serial No. 2092, 1997, p. 30.
4. Bijen, J. Benefits of Slag and Fly Ash. *Construction and Building Materials*, Vol. 10, No. 5, 1996, pp. 309–314.
5. Boddy, A. M., R. D. Hooton, and M. D. A. Thomas. The Effect of Product Form of Silica Fume on Its Ability to Control Alkali–Silica Reaction. *Cement and Concrete Research*, Vol. 30, No. 7, 2000, pp. 1139–1150.

6. Shao, Y., T. Lefort, S. Moras, and D. Rodriguez. Studies on Concrete Containing Ground Waste Glass. *Cement and Concrete Research*, Vol. 30, No. 1, 2000, pp. 91–100.
7. Dyer, T. D., and K. D. Ravindra. Chemical Reactions of Glass Cullet Used as Cement Component. *Journal of Materials in Civil Engineering*, Vol. 13, No. 6, 2001, pp. 412–417.
8. Shayan, A., and A. Xu. Value-Added Utilisation of Waste Glass in Concrete. *Cement and Concrete Research*, Vol. 34, No. 1, 2004, pp. 81–89.
9. Afshinnia, K., and P. R. Rangaraju. Efficiency of Ternary Blends Containing Fine Glass Powder in Mitigating Alkali–Silica Reaction. *Construction and Building Materials*, Vol. 100, 2015, pp. 234–245.
10. Chen, C. H., R. Huang, J. K. Wu, and C. C. Yang. Waste E-Glass Particles Used in Cementitious Mixtures. *Cement and Concrete Research*, Vol. 36, No. 3, 2006, pp. 449–456.
11. Rangaraju P.R., H. Rashidian-Dezfouli, G. Nameni, and G.Q. Amekuedi. Properties and Performance of Ground Glass Fiber as a Pozzolan in Portland Cement Concrete. Proceedings of 2016 International Concrete Sustainability Conference, Washington DC, May 15-18, 2016, p. 10.
12. Rashidian-Dezfouli, H., and P. R. Rangaraju. Role of Ground Glass Fiber as a Pozzolan in Portland Cement Concrete. *Transportation Research Record: Journal of the Transportation Research Board*, 2017. 2629: 33–41.
13. Kamali, M., and A. Ghahremaninezhad. An Investigation into the Hydration and Microstructure of Cement Pastes Modified with Glass Powders. *Construction and Building Materials*, Vol. 112, 2016, pp. 915–924.
14. Taylor, H. F. W. *Cement Chemistry*. Thomas Telford, London, 1997.
15. Stade, H. On the Reaction of CSH (di, poly) with Alkali Hydroxides. *Cement and Concrete Research*, Vol. 19, No. 5, 1989, pp. 802–810.
16. Duchesne, J., and M. Bérubé. Long-Term Effectiveness of Supplementary Cementing Materials against Alkali–Silica Reaction. *Cement and Concrete Research,* Vol. 31, No. 7, 2001, pp. 1057–1063.
17. Ramlochan, T., P. Zacarias, M. D. A. Thomas, and R. D. Hooton. The Effect of Pozzolans and Slag on the Expansion of Mortars Cured at Elevated Temperature: Part I: Expansive Behaviour. *Cement and Concrete Research*, Vol. 33, No. 6, 2003, pp. 807–814.
18. Hong, S. Y., and F. P. Glasser. Alkali Sorption by CSH and CASH Gels: Part II. Role of Alumina. *Cement and Concrete Research*, Vol. 32, No. 7, 2002, pp. 1101–1111.
19. Chappex, T., and K. L. Scrivener. The Influence of Aluminium on the Dissolution of Amorphous Silica and Its Relation to Alkali Silica Reaction. *Cement and Concrete Research*, Vol. 42, No. 12, 2012, pp. 1645–1649.
20. Marchand, J., I. Odler, and J. P. Skalny. *Sulfate Attack on Concrete*. CRC Press, New York, 2003.
21. Tian, B., and M. D. Cohen. Does Gypsum Formation during Sulfate Attack on Concrete Lead to Expansion? *Cement and Concrete Research*, Vol. 30, No. 1, 2000, pp. 117–123.

The Standing Committee on Concrete Materials and Placement Techniques (AFN40) peer-reviewed this paper (18-05767).

Article

TRR
JOURNAL OF THE TRANSPORTATION RESEARCH BOARD

Transportation Research Record
2018, Vol. 2672(27) 99–108

DOI: 10.1177/0361198118801357
journals.sagepub.com/home/trr
SAGE

Sustainability of Using Recycled Concrete Aggregates as Coarse Aggregate in Concrete Pavements

Farhad Reza[1], W. James Wilde[1], and Bernard Izevbekhai[2]

Abstract

Conservation and reuse of resources are important aspects of sustainability. It is common practice in the U.S. to crush old concrete pavements at the end of their service lives into recycled concrete aggregate and then reuse them in the base course of new pavements. It is not, however, common to use the recycled concrete aggregate (RCA) in the new concrete pavement itself. This paper examines the sustainability and, in particular, the economics of using RCA in the construction of new concrete pavements. Life cycle cost analyses were performed for various hypothetical pavement construction scenarios including RCA both in the base and in the concrete layer and compared with conventional options. If RCA is simply substituted for natural aggregate in a concrete mixture, the concrete is likely not to perform as well as a concrete made without RCA. This can be compensated for by decreasing the water–cement ratio and/or increasing the cement content. Despite this fact, the results of the life cycle cost analyses show that it can be very economical to use RCA in new concrete pavement construction. As the natural aggregate used for concrete may be more expensive than the natural aggregate used for base courses, it may make good economic sense to substitute RCA for natural aggregate in concrete. Life cycle analysis also shows that concrete with RCA may have better environmental impact.

Sustainability practices dictate that, whenever possible, materials should be recycled and reused. In fact, the American Society of Civil Engineers Code of Ethics Canon 1 requires civil engineers to consider sustainability in any design. In a 2004 study conducted by the Federal Highway Administration (FHWA), it was reported that 38 states in the U.S. use recycled concrete aggregate (RCA) as aggregate base, but only 11 had used RCA in Portland cement concrete (PCC) (*1*). In the 1980s, several states built trial projects using RCA in PCC pavements, most of which were built in Michigan, Wisconsin, Iowa, and Minnesota (*2*). Michigan subsequently placed a moratorium on the use of RCA in PCC as a result of the observation of some mid-slab transverse cracking in the trial projects, although subsequent studies attributed the cracking to flaws in the design (such as long joint spacing) rather than material-related distress. Some of the other states reported relatively good performance of the RCA pavements. Since that initial period of activity, not much new construction has taken place. In a 2011 study conducted by Iowa State University, some of the perceived barriers by state highway agencies preventing widespread implementation of RCA in PCC pavements were assessed, and the authors concluded that most of the problems had solutions already in the literature and thus it was likely a technology transfer problem (*3*). More recently, two pavements were created with the two-lift construction procedure with RCA being used in the lower layer. One of these projects was built at the MnROAD test facility in 2010 (*4*) and the other at the Illinois Tollway construction in 2013–2014. This paper examines the sustainability, and in particular, the economics of using RCA in the construction of new concrete pavements. Life cycle cost analyses (LCCA) were performed for various hypothetical pavement construction scenarios including RCA both in the base and in the concrete layer and compared with conventional options. Life cycle analysis (LCA) was also performed to compare environmental impacts such as energy consumption and emissions of a concrete mix with 100% substitution of RCA for coarse aggregate versus a conventional concrete mix.

[1]Department of Mechanical & Civil Engineering, Minnesota State University, Mankato, MN
[2]MnDOT Office of Materials and Road Research, Maplewood, MN

Corresponding Author:
Address correspondence to Farhad Reza: farhad.reza@mnsu.edu

Literature Review

The mechanical properties of concrete made with RCA have been investigated by various researchers. Most of the literature suggests limiting or eliminating the use of RCA fines in new concrete because of workability and durability issues (*2*). The compressive strength of concrete made using coarse RCA is typically lower (up to 20%) than that of concrete with natural aggregates (*2*); however, the strength could be even lower when the new concrete has a lower water–cement ratio because of the weaker interfacial transition zone of the RCA (*5*). Flexural strengths have also been found to be lower by up to 10% (*2*). Hardened concrete with RCA generally has 20–40% lower modulus of elasticity than that of conventional concrete, and 10–30% higher coefficient of thermal expansion (*6*, *7*).

Studies have reported increase freeze–thaw resistance of concrete made with RCA because of the higher entrained air contents that result from the air entrainment contained in the RCA aggregate (*2*). In general, the D-cracking susceptibility of RCA is less because the susceptible aggregate is crushed to a smaller size in producing the RCA (*2*). Alkali silica reaction potential is higher in mixtures that use RCA because more aggregate surfaces are exposed for the reaction by the crushing operation (*2*). This can be mitigated by using low-alkali, Type II cement, blending the RCA with quality conventional aggregates, and using fly ash in the mix. Lower quality concrete may result in RCA with low abrasion resistance. The American Association of State Highway Transportation Officials recommends a limit of 50% mass loss by the Los Angeles Rattler for RCA to be used in concrete (*8*).

Donalson et al. performed a sustainability assessment using RCA versus virgin limestone aggregate (VLA) in base courses from the perspective of environmental, social, and economic aspects (*9*). The environmental impact was found to demonstrate a reduced impact in favor of RCA in process energy and disposal (viz. fuel used to haul concrete to landfill) at −0.01 and + 0.04 metric tons of carbon dioxide equivalents, respectively. The delivered price of the material was found to differ primarily in the transportation costs, as the processing procedures were found to be similar. For a hypothetical scenario in Florida, the delivered cost was $12.75 per ton for RCA and $13.20 per ton for VLA (based on a 30-mile haul distance for the VLA). For the social impact, the leachability of several chemicals for both RCA and VLA was evaluated. It was found that RCA leaches less harmful and less toxic amounts of chemicals than VLA; however, in both cases the actual amount of constituent reaching the groundwater is negligible compared with groundwater contamination limits of the Environmental Protection Agency.

Ram et al., in a report developed for the Michigan Department of Transportation, performed LCCA to evaluate the use of RCA in concrete pavements (*10*). Pavements constructed using RCA exhibited comparable performance (in terms of life cycle agency costs) to sections with natural coarse aggregates at lower traffic volumes; however, at higher traffic volumes the majority of them underwent major rehabilitation or reconstruction activities after about 20 years of service. Michigan's initial trials of RCA pavements were often jointed reinforced concrete pavement and suffered mid-slab transverse cracking that may have been caused by some combination of small top size aggregate, poor aggregate interlock owing to adhered mortar, long joint spacing, and too little reinforcement. The average agency cost for level I (<6000 annual average daily traffic (AADT), ~9 inch average thickness) pavements was almost the same at about $15,000 on an annual worth basis for both natural aggregate and RCA concrete for a 50-year analysis period. For level III (>10,000 AADT, ~11 inch thickness), pavements the average agency cost was about $14,600 for natural aggregate and about $12,170 for RCA. Thus, the agency costs seem to be approximately the same or slightly better for RCA when evaluated over a longer period of time, even after considering the 20-year service life of some of their RCA pavements.

Horvath developed a LCA framework in Microsoft Excel for the Recycled Material Resource Center (*11*). The tool name is Pavement Life Cycle Assessment Tool for Environmental and Economic Effects (PaLATE). A user-guide for this tool was developed as part of a Master's Thesis by Nathman (*12*). PaLATE assesses the environmental and economic effects of pavement and road construction. Users input initial designs, initial construction materials, maintenance materials and processes, equipment, and costs for a project. Environmental outputs include energy consumption; water consumption; emissions of carbon dioxide, nitrogen oxide, particulate matter less than 10 microns, sulfur dioxide, and carbon monoxide; and leachate information (including mercury, lead, Resource Conservation and Recovery Act [RCRA] hazardous waste, and both cancerous and non-cancerous human toxicity potential).

Verian et al. performed a benefit–cost analysis and developed a useful spreadsheet to evaluate initial material costs associated with RCA use for the Indiana DOT (*13*). For the base aggregate, two different gradations were considered: #8 and #53. When concrete is crushed into aggregate, there is about 50% waste (i.e., fines) for the #8 and 40% waste for the #53 gradation. Some other assumed values in the analysis included:

- $5/ton RCA crushing costs
- $9.50/ton concrete natural aggregate

- $9.50/ton base #8
- $9/ton base #53
- $ 0.35/ton/mile hauling cost
- $2/ton landfill cost

The haul distances assumed were:

- Old project to crushing operation: 10 miles
- Crushing to delivery point: 7.4 miles
- Project to landfill: 50 miles
- Natural aggregate source to delivery point: 21 miles

Under these assumptions, for 50% coarse RCA in the concrete and 100% RCA in the base, they achieve a net cost savings of about $2.50/ton of aggregate.

Hu et al. performed a review of two-lift construction in the U.S. (*14*). A survey of contractors showed that most would favor using a second batch plant instead of adding bins for the first plant. In addition, a second slip-form paver is required for the wet–wet method of construction. Despite the increased equipment and labor costs, the three most recent two-lift construction projects in the U.S. (Kansas 2008, Minnesota 2010, and Illinois 2012) showed a net saving compared with the conventional concrete alternative. The positive data result from the substantial saving from the reduced aggregate costs and concrete cost used in the bottom lift, together with the positive bidding climate currently seen in the U.S. The Minnesota project consisting of MnROAD Cells 71 and 72 on the I-94 mainline showed concrete costs of $71.07/cy concrete costs, $2.61/sy paving costs, and $20.38/sy net costs for the conventional concrete compared with $62.66/cy, $4.28/sy paving costs, and $19.94/sy net costs for the two-lift construction.

Objectives

The fact that numerous researchers have shown that the mechanical properties of RCA are inferior to those of natural aggregates indicates that the long-term performance of pavements made with RCA would be expected to be worse than that of pavements made with natural aggregate when the two are made with similar structural design and concrete mix design. This assumption has been corroborated by comparing the long-term performance of pavements made with RCA versus those made with natural aggregate (*15*). To circumvent this problem, changes could be made at the design stage, such as using a thicker pavement or making a stonger concrete mix; however, these would increase cost. The main objective of this paper was to examine if the use of RCA in concrete pavements could still be economically competitive. This was done through LCCA for various hypothetical pavement construction scenarios including RCA both in the base and in the concrete layer and compared with conventional options. LCA was also performed to compare environmental impacts such as energy consumption and emissions of a concrete mix with 100% substitution of RCA for coarse aggregate versus a conventional concrete mix.

Life Cycle Cost Analysis

LCCA is a widely used technique in the U.S. for comparing pavement alternatives. From engineering economics theory, alternatives can be compared equivalently by future worth, present worth, or annual worth. In LCCA for pavement alternatives, it is most common to compare present worth (or net present value) or annual worth (or effective uniform annual cost). It is useful to review the relationships between some of these values as given by Equations 1 through 3:

$$F = P(1 + i)^n$$

F = Future worth; P = Present worth (1)

i = Annual real discount rate; n = number of years

$$P = \frac{F}{(1 + i)^n} \quad (2)$$

$$A = P\left[\frac{i(1 + i)^n}{(1 + i)^n - 1}\right] \quad (3)$$

A = Annual worth

The real discount rate is a function of both the interest rate and inflation rate and is given by:

$$i = \frac{1 + i_{int}}{1 + i_{inf}} - 1 \quad (4)$$

In this study, a 1.5% discount rate was used, which is the current recommended discount rate for 30 years by the Office of Management and Budget (OMB). It is suggested by OMB that for analysis periods longer than 30 years, the 30-year value should still be used.

The LCCA were performed for concrete pavement construction using conventional concrete, concrete with 50% RCA replacement for coarse aggregate, and concrete with 100% RCA replacement for coarse aggregate. For additional comparisons, the concrete mix design that was used in the two-lift construction at the MnROAD test facility was also evaluated.

The authors evaluated concrete mixtures made at a water–cementitious (w/cm) material ratio of 0.37 with 0%, 50%, and 100% volume replacement of RCA for natural coarse aggregate (*15*). The results from some of the experiments with respect to the control for the 50% and 100% RCA replacement respectively were—compressive strength: 32% lower and 43% lower, flexural strength: 17% lower and 20% lower, drying shrinkage: 1% higher and 32% higher, and coefficient of thermal expansion: 1% higher and 11% higher. As the mechanical properties of concrete containing RCA tend to be worse than those for conventional concrete when the RCA is simply substituted for natural coarse aggregate, it can be reasonably expected that the performance of pavements with RCA would be worse than that of the conventional concrete. Indeed, an analysis of the historical pavement performance data for RCA pavements built in Minnesota in the 1980s compared with similar conventional concrete pavements constructed in the same period and subjected to a similar range of traffic showed that the non-RCA pavements had about 18.5% more longevity [defined here as the time taken for the pavement to reach a condition where a major concrete pavement rehabilitation (CPR) is required] than the RCA pavements (*15*).

To compensate for this difference in performance between RCA and non-RCA pavements, assumed to be caused by the difference in mechanical properties, two possible techniques would be to (a) modify the concrete mix design to achieve a stronger, more durable concrete, or (b) increase the pavement thickness to decrease applied stresses. Under option (a), additional experiments were conducted to obtain suitable concrete mix designs that could achieve similar mechanical properties to the control mix. This was achieved by using lower w/cm ratios and increased cement contents. For the case of 50% RCA replacement, satisfactory performance was found at w/cm of about 0.36, and for the case of 100% replacement, satisfactory performance was found at w/cm of about 0.35.

Under option (b), the pavement thickness could be increased to achieve similar performance with the conventional concrete. While there is a lot of literature concerning static properties of concrete containing RCA, there is relatively little research done on the fatigue properties. In one study investigating the flexural fatigue of concrete containing RCA, the authors found that the 2 million cycle endurance limit for stress level S was 0.58 for normal aggregate concrete, 0.56 for 50% RCA concrete, and 0.50 for 100% RCA concrete (*16*). The stress level S is the ratio of the maximum tensile stress induced in the pavement to the flexural strength or modulus of rupture of the pavement. With the stress level and flexural strength known, the maximum bending stress can be calculated. From classical bending theories, the stress σ varies inversely with the thickness squared:

$$\sigma = \frac{Mc}{I} = \frac{M(h/2)}{bh^3/12} = \frac{6M}{bh^2} \qquad (5)$$

where M is the bending moment, c is the distance from the neutral axis, I is the area moment of inertia, b is the width of the beam, and h is the thickness of the pavement. For any value of M and b, the required thickness of the different concrete mix designs normalized with respect to the control mix can be obtained. Thus, it was determined that to achieve the same fatigue performance life, the thickness of the 50% RCA concrete pavement should be about 12% greater than the control mix pavement, whereas the 100% RCA pavement should be about 24% thicker.

It is normally suggested to use a relatively long analysis period to be able to capture the effects of various maintenance activities. In this study, the analysis period was chosen to be 50 years, which is a reasonable timeline considering the long expected service lives of new concrete pavements in Minnesota. A reasonable expected time to major CPR of 50 years has been assumed for 50% RCA pavements, and the time to major CPR for non-RCA pavements has been taken as 59 years, (about 18.5% longer), following the trend seen in the historical pavement performance in Minnesota. The 100% RCA pavement's expected time to major CPR has been taken as 46 years essentially using a proportional value of flexural strength compared with the 50% RCA concrete, thus, a major CPR would occur during the analysis period for the 100% RCA concrete.

One of the major differences between the alternatives considered was the concrete mix design. While the unit price of concrete as a whole (e.g., in $ per cy or $ per sy) is normally reported in bids and is readily available in MnDOT average bid prices, it does not allow the nuances of changes in mix design to be accounted for in the LCCA. For example, the difference in cost for RCA versus natural aggregate within the mix cannot be evaluated with the lump sum value of concrete, and thus the mix must be broken down into its components.

Additional initial costs considered were joint sawcuts and dowel bars. These needed to be included because some of the alternatives called for a 12 ft joint spacing instead of 15 ft standard joint spacing. The 100% RCA concrete at 0.37 w/cm had a much higher drying shrinkage and coefficient of thermal expansion than that of the control mix. Therefore, it was recommended to use a 12 ft spacing instead of 15 ft for this mix.

It is also realized that there is an effect of "economies of scale." For example, in the MnDOT average bid prices, it is often found that unit costs are lower for larger quantities. For this reason, a minimum length of 7 miles for a new paving project was assumed in this analysis. Annual maintenance costs have been assumed to be the same for all alternatives and were not included in the analysis.

Table 1. Unit Costs Assumed in LCCA

Material/Process	Unit cost	Unit	Reference
RCA	$8.75	ton	Izevbekhai (2015) (*18*)
Natural aggregate for concrete	$22.00	ton	Sibley Aggregate, Kraemer Mining MN (2016)
Aggregate for EAC	$20	ton	Izevbekhai (2015) (*18*)
Natural aggregate for base class 5	$16.77	ton	MNDOT average bid (2014) (*19*)
Sand	$5	ton	PaLATE, producers websites
Cement	$110	ton	PaLATE, Cemstone (2016)
Flyash	$90	ton	PaLATE, Cemstone (2016)
Water	$36.88	Mgal	MNDOT average bid (2014) (*19*)
High range water reducer	$1000	ton	Internet search (2016)
Haulage	$ 0.35	ton/mile	Verian et al. (2013) (*13*)
Sawcut	$4.83	LF	MNDOT average bid (2014) (*19*)
Dowels	$6.18	ea	MNDOT average bid (2014) (*19*)
Landfill fee	$4	ton	Midwest Asphalt (2016)
Premium for 2-Lift	$1.66	SY	MnROAD Test Cell (Hu et al.) (*14*)
Diamond grinding	$2.30	SY	MNDOT average bid (2014) (*19*)

During the study of the historical pavement performance data it was found that the frequency of minor CPR for both RCA and non-RCA pavements were fairly similar, and thus they have not been included in this analysis. One of the alternatives was assumed to have a major CPR at 46 years. The CPR technique has been assumed to be diamond grinding. The future cost of this operation can be converted to present worth using Equation 2 and to annual worth using Equation 3. In the case that any of the alternatives have some remaining service life at the end of the analysis period, one of the techniques to account for this is to calculate the residual value. The residual value is given by:

$$\text{Residual Value} = \frac{\text{Remaining Service Life}}{\text{Expected Service Life}} \times \text{Initial Cost} \tag{6}$$

The residual value can then be converted to present worth and annual worth. It should be noted that when comparing alternatives with LCCA, any costs that are constant between the alternatives need not be considered. In fact, if the constant costs *are* included, it will tend to lessen the differences between the various alternatives considered. As mentioned, annual maintenance and minor CPR have not been included. Similarly, the landfill disposal of non-usable RCA (fines) would be constant between all the alternatives and has not been calculated. Using data from Verian et al. (*13*), the amount of non-usable RCA (fines) has been taken as 40%.

The old Portland cement concrete pavement (PCCP) to be recycled has been assumed to be 7 miles long by 24 ft wide by 10 in thick. The base of the new PCCP to be constructed has been assumed to be 7 miles by 24 ft wide by 8 in thick class 5 aggregate. All of the new PCCP alternatives are 7 miles long and 24 ft wide. The hauling distance between the natural aggregate sources and the delivery point has been assumed to be 30 miles.

Table 2. Unit Weights of Materials Used

Material	Unit weight, ton/cy
Natural aggregate for concrete	2.33
Natural aggregate for class 5 base	2.33
Aggregate for exposed aggregate concrete	2.33
Natural fine aggregate	2.21
Recycled coarse aggregate	2.00

The new conventional concrete pavement (Alternative 1) had a thickness of 9 in. As discussed previously, it was suggested that in order to achieve similar fatigue performance as the conventional concrete, pavement thickness be increased by 12% for 50% RCA concrete and 24% for 100% RCA concrete. In lieu of increasing the thickness, the concrete mix could also be strengthened by increasing the cementitious material content and decreasing the water–cementitious (w/cm) ratio. Based on the laboratory testing, it is suggested that the w/cm ratio be decreased to 0.36 for 50% RCA concrete and 0.35 for 100% RCA concrete to achieve similar fatigue performance as the conventional concrete (with w/cm of 0.37). Also, because of higher drying shrinkage and higher coefficient of thermal expansion, it was recommended to reduce the joint spacing from 15 ft to 12 ft for 100% RCA concrete at the w/cm of 0.37; however, at w/cm of 0.35 the shrinkage and thermal expansion of the 100% RCA mix were similar to that of the conventional concrete, and therefore the joint spacing could remain at the standard 15 ft value.

The assumed unit costs for the LCCA calculations are given in Table 1 and the unit weights for some of the materials are given in Table 2. The concrete mix designs

Table 3. Concrete Mix Proportions (lb/cy)

Alternative in which used	1	2 & 3	4	5 & 6	7	8	
Mix description	0% RCA	50% RCA, 0.37 w/cm	50% RCA, 0.36 w/cm	100% RCA, 0.37 w/cm	100% RCA, 0.35 w/cm	Lower lift mix	Top lift mix
Cement	410	410	417	410	420	360	616
Fly ash	175	175	178	175	182	240	109
Water	216	217	214	216	211	234	283
Natural coarse aggregate	1819	909	909	0	909	825	1976
Natural fine aggregate	1309	1309	1309	1309	1309	1200	843
Recycled coarse aggregate	0	780	780	1560	1560	920	0
High range water reducer	0.83	1.38	3.03	1.66	5.64	0.88	0.88
w/cm	0.37	0.37	0.36	0.37	0.35	0.39	0.39

utilized in this study are summarized in Table 3. The two mixes for the two-lift construction were from the MnROAD test cell mixes (*17*).

A total of eight alternatives were considered in the LCCA, namely:

1. Conventional concrete pavement with no RCA in the concrete with an expected life until first major CPR of 59 years.
2. 50% coarse RCA in the concrete with similar structural design as (1), but with an expected life to first major CPR reduced to 50 years.
3. 50% coarse RCA in the concrete with a 12% thicker design than (1) and similar life expectancy of 59 years to first major CPR.
4. 50% coarse RCA in the concrete with a lower w/cm ratio (0.36) and similar structural design as (1) with a similar life expectancy of 59 years to first major CPR.
5. 100% coarse RCA with similar structural design as (1) but with reduced joint spacing of 12 ft and reduced life to first major CPR of 46 years.
6. 100% coarse RCA in the concrete with a 24% increase in thickness from (1) and reduced joint spacing of 12 ft and similar life expectancy of 59 years to first major CPR.
7. 100% coarse RCA in the concrete with a lower w/cm ratio (0.35) and similar structural design as (1) with a similar life expectancy of 59 years to first major CPR.
8. Two-lift construction that has 50% RCA in the lower lift and exposed aggregate concrete in the top lift with a similar life expectancy to first major CPR as (1), that is, 59 years.

Round 1 of Analysis

The practice of utilizing crushed RCA in the base is common in Minnesota; however, there may not be enough quantity of RCA obtained locally on a project to satisfy the demand for the base course (up to 75% of the blend can be RCA according to MnDOT Specifications) let alone in the concrete. In round 1 of the analysis, the priority usage of the locally crushed RCA was in the concrete pavement. If there was any RCA remaining, it was then applied to the base course.

Round 2 of Analysis

A review of several aggregate suppliers in Minnesota revealed that class 5 recycled concrete base could be purchased at a very low cost (average of about $5.25 per ton). This is significantly less than the $16.77 per ton for natural class 5 base (MnDOT average bid price). In light of this fact, a second round of LCCA was performed assuming that the contractor would purchase enough extra RCA base aggregate to make all of the bases using 75% recycled and 25% natural aggregates.

Results

An example of the analysis for the 100% RCA mix (Alternative 7) is shown in Tables 4 and 5. The results of the first round of LCCA are presented in Table 6 in increasing order of costs from left to right. The results clearly indicate that the use of RCA in concrete can be economical, with two of the alternatives being less costly than the conventional concrete alternative. It can be seen that pavement longevity is valuable, with those alternatives with lower expected lives to the first major CPR being among the most expensive. When evaluating the option of improving life expectancy either by thickening the pavement slab or modifying the concrete mix to a stronger design, the latter proved to be more economical. The most economical alternative overall was number 7, which was the 100% RCA mix with a lower w/cm ratio. Although this alternative includes an increased cementitious content which is an expensive component of the concrete, the benefit achieved by substituting the less

Table 4. Mix Analysis—Alternative 7 (100% RCA, 15 ft Spacing, Reduced w/cm)

Variable	Length, mi	Width, ft	Thickness, in	Area, sy	Volume, cy	Usable %	Waste	Usable vol, cy	Waste vol, cy
Old PCCP	7	24	10	98,560	27,378	60%	40%	16,427	10,951
New PCCP	7	24	9	98,560	24,640				
						Vol. of remaining RCA, cy	% Recycled material	% Natural material	
Base	7	24	8	98,560	21,902	6,810	31%	69%	
Concrete mix components		Weight, pcy	Volume, cy		Total project volume, cy		Total weight for project, tons	Cost	Concrete cost, $/cy
Cement		420	0.08		1,950		5,174	$214,498	
Fly ash		182	0.04		986		2,242	$201,802	
Water		211	0.13		3,081		2,600	$23.0	
Natural coarse aggregate		0	0.00		0		0	$ 0	
Natural fine aggregate		1,309	0.30		7,299		16,126	$80,629	
Recycled coarse aggregate		1,560	0.39		9,617		19,214	$168,122	
HRWR		5.64	0.003		81		69	$69,485	
									$29.81

Table 5. Economic Analysis—Alternative 7 (100% RCA, 15 ft Spacing, Reduced w/cm)

Hauling distances and costs	Distance, mi	Cost, $
Crushing old concrete	0	$ 0
Delivery of RCA	0	$ 0
Delivery of natural aggregate	30	$538,555
Initial costs	Total cost, $	Unit cost, $/sy
Concrete	$665,074	
Recycled portion of base	$119,173	
Natural portion of base	$589,720	
Haulage	$538,555	
Sawcutting	$285,627	
Dowels	$365,460	
Total initial cost	$2,563,609	$26.01
Analysis results		per sy
Expected time to Major CPR, yrs	59	
Analysis period, yrs	50	
Remaining service life, yrs	9	
Residual value, $	$461,450	$4.68
Present value of Residual value, $	$219,191	$2.22
Cost of major CPR, $	$ 0	$ 0
Present value of major CPR, $	$ 0	$ 0
Total NPV, $	$2,344,419	$23.79
Equivalent uniform Annual cost, $	$66,984	$ 0.68

expensive RCA for the natural aggregate outweighs this. As natural aggregate costs for concrete tend to be higher than the natural aggregate costs for bases, it seems to make good economic sense to utilize the locally available crushed RCA from a project first in the concrete rather than the base course. Alternatives 6 and 5, both with 100% RCA, are adversely impacted with the need to reduce joint spacing from 15 ft to 12 ft, and the correspondingly higher costs for sawcuts and dowel bars. The two-lift construction finished somewhere near the middle, demonstrating that savings in material costs can help to offset increased labor and equipment costs. It should be noted, however, that there is not enough historical data to have much knowledge on the long-term performance, predicted service life, and maintenance activities for the two-lift pavement.

As noted previously, it was found that RCA which would meet the class 5 gradation for base was available from several aggregate suppliers at a substantially lower price than natural aggregate for class 5 base. The second round of analysis was performed assuming the contractor would purchase extra RCA from offsite to achieve the maximum allowable recycled content of 75% recycled in the base. The results of this analysis are shown in Table 7 with the cost increasing from left to right. All of the costs are significantly reduced compared with Table 6. Under this scenario, the fact that RCA use in concrete can be highly economical is even more pronounced, with most alternatives proving less costly than the conventional concrete pavement (alternative 1). The most economical

Table 6. Results of LCCA for Round 1

Variable	Alternative 7	4	1	3	8	2	6	5
NPV, millions $	$2.344	$2.393	$2.440	$2.531	$2.606	$2.717	$2.759	$2.821
NPV, $/sy	$23.79	$24.27	$24.76	$25.68	$26.44	$27.57	$27.99	$28.62
Annual cost, $	$66,984	$68,358	$69,713	$72,326	$74,447	$77,629	$78,829	$80,606
Annual cost ($/sy)	$ 0.68	$ 0.69	$ 0.71	$ 0.73	$ 0.76	$ 0.79	$ 0.80	$ 0.82
Assumed life to 1st Major CPR, yrs	59	59	59	59	59	50	59	46
Thickness, in	9	9	9	10.08	9	9	11.16	9
Joint spacing, ft	15	15	15	15	15	15	12	12
RCA in concrete, %	100	50	0	50	31	50	100	100
RCA in base, %	31	53	75	50	59	53	21	31
Concrete cost, $/cy	$29.81	$34.85	$40.07	$33.75	$19.91	$33.75	$27.30	$27.30

alternatives were unchanged from round 1, that is, Alternatives 7 and 4. The trends seen previously also held here. When looking at the alternatives containing RCA, the most expensive ones tend to be those with shorter life expectancy to first major CPR. It appears to be cheaper to change the mix design to a stronger, more durable mix than to increase the thickness of the pavement.

Life Cycle Analysis

The triple bottom line principles by which sustainability can be measured are economic, environmental, and social benefits. According to the FHWA, the goal of sustainability is the satisfaction of basic social and economic needs, both present and future, and the responsible use of natural resources, all while maintaining or improving the well-being of the environment on which life depends. A sustainable highway should satisfy life cycle functional requirements of societal development and economic growth while reducing negative impacts to the environment and consumption of natural resources.

LCA was performed with the PaLATE tool. A comparison of Alternative 7 (100% RCA mix) against Alternative 1 (conventional concrete) was performed in PaLATE. The results for Alternatives 7 and 1 are given in Tables 8 and 9, respectively. The output includes several parameters such as energy consumption, water consumption, various emissions, potential leachates, and human toxicity. It can be seen that in most cases the values are fairly similar and in many cases the 100% RCA mix appears to have more environmental benefit.

Conclusion

An economic analysis using the LCCA approach shows that it can be very economical to use RCA in new concrete pavement construction. The practice of utilizing crushed RCA in the base course is already quite common in Minnesota; however, because the natural aggregate for concrete may be more expensive than that for base courses it may make good economic sense to substitute RCA for natural aggregate in concrete.

Long-life pavements tend to be more economical than shorter-life pavements (with lower initial costs) in the long run. Concrete pavements made with RCA may have shorter life spans than those of their conventional concrete counterparts at the same w/cm ratio. To achieve similar performance, one option may be to change the structural design (e.g., increase the PCCP thickness) and another option may be to strengthen the mix (e.g., increase the cementitious material and decrease the w/cm ratio). The LCCA presented in this paper tends to favor the option of strengthening the concrete mix.

Table 7. Results of LCCA for Round 2 (Extra RCA Purchased for Base)

	Alternative							
Variable	7	4	6	3	5	8	1	2
NPV, million $	$1.878	$2.159	$2.181	$2.270	$2.311	$2.431	$2.440	$2.462
NPV, $/sy	$19.05	$21.91	$22.12	$23.03	$23.45	$24.66	$24.76	$24.98
Annual cost, $	$53,657	$61,695	$62,303	$64,862	$66,033	$69,454	$69,713	$70,342
Annual cost, $/sy	$ 0.54	$ 0.63	$ 0.63	$ 0.66	$ 0.67	$ 0.70	$ 0.71	$ 0.71
RCA in concrete, %	100	50	100	50	100	31	0	50
RCA in base, %	75	75	75	75	75	75	75	75

Table 8. Alternative 7 (100% RCA Mix) LCA Results from PaLATE

Variable	Materials production	Materials transportation	Processes (equipment)	Total
Energy (MJ)	42.8×10^6	2.34×10^6	428×10^3	45.6×10^6
Water consumption (kg)	15,900	399	42	16,300
CO_2 (mg) = GWP	3,200	175	32	3,400
NO_x (kg)	36,900	9,400	677	47,000
PM_{10} (kg)	16,900	1,800	48	18,800
SO_2 (kg)	24,100	565	45	24,800
CO (kg)	19,600	785	146	20,600
Hg (g)	60	2	0	62
Pb (g)	3,700	79	14	3,800
RCRA hazardous waste generated (kg)	65,300	16,900	3,100	85,300
Human toxicity potential (cancer)	1.45×10^6	50,300	0	1.5×10^6
Human toxicity potential (non-cancer)	6.5×10^9	61.7×10^6	0	6.5×10^9

Table 9. Alternative 1 (Conventional Mix) LCA Results from PaLATE

Variable	Materials production	Materials transportation	Processes (equipment)	Total
Energy (MJ)	42.8×10^6	2.41×10^6	423×10^3	45.6×10^6
Water consumption (kg)	15,900	411	41	16,300
CO_2 (mg) = GWP	3,000	180	32	3,200
NO_x (kg)	36,500	9,690	669	46,900
PM_{10} (kg)	17,200	1,890	48	19,200
SO_2 (kg)	24,000	581	44	24,600
CO (kg)	19,600	808	144	20,500
Hg (g)	60	2	0	62
Pb (g)	3,660	81	14	3,800
RCRA hazardous waste generated (kg)	64,500	17,400	3,050	86,000
Human toxicity potential (cancer)	1.76×10^6	51,700	0	1.81×10^6
Human toxicity potential (non-cancer)	10.4×10^9	63.5×10^6	0	10.5×10^9

In most cases, the amount of RCA available from crushing an existing PCCP locally is expected to be less than what is needed to construct both the new concrete pavement as well as the base course. At current rates, the cost of purchasing recycled concrete base is significantly less than that of natural aggregates; therefore, if the contractor purchases extra RCA to make the entire base up to 75% recycled, which is the maximum allowed in Minnesota, then the project can become even more economical.

Economics is just one of the triple bottom line principles of sustainability, the others being environment and social aspects. Environmental and social impacts were also evaluated utilizing the LCA framework in the PaLATE program. The use of a large amount of RCA in the concrete had several advantages over the conventional concrete, for example in decreased human toxicity potential, reduced energy consumption, and some reduced emissions.

Acknowledgments

This research work was funded by the Minnesota Department of Transportation under Contract No. 05309.

Author Contributions

The authors confirm contribution to the paper as follows: study conception and design: FR, WJW, BII; data collection: FR, WJW, BII; analysis and interpretation of results: FR, WJW, BII; draft manuscript preparation: FR, WJW, BII. All authors reviewed the results and approved the final version of the manuscript.

References

1. Federal Highway Administration. *Transportation Applications of Recycled Concrete Aggregate*. FHWA State of the Practice National Review, FHWA, Washington, D.C., 2004.
2. Anderson, K. W., J. S. Uhlmeyer, and M. Russell. *Use of Recycled Concrete Aggregate in PCCP: Literature Search*. Report WA-RD 726.1. Washington Department of Transportation, Olympia, WA, 2009.
3. Garber, S., R. Rasmussen, T. Cackler, P. Taylor, D. Harrington, G. Fick, M. Snyder, T. Van Dam, and C. Lobo. *Development of a Technology Deployment Plan for the Use of Recycled Concrete Aggregate in Concrete Paving Mixtures*. FHWA DTFH61-06-H-00011. Iowa State University, Ames, IA, 2011.
4. Tomkins, D., L. Khazanovich, M. Vancura, M. I. Darter, and S. Rao. Construction of Sustainable Pavements: Two-Layer Concrete Pavements at the MNROAD Facility. Presented at 90th Annual Meeting of the Transportation Research Board, Washington, D.C., 2011.
5. Otsuki, N., S. Miyazato, and W. Yodsudjai. Influence of Recycled Aggregate on Interfacial Transition Zone, Strength, Chloride Penetration and Carbonation of Concrete. *Journal of Materials in Civil Engineering*, Vol. 15, No. 5, 2003, pp. 443–451.
6. Burke, T., M. Cohen, and C. Scholer. *Synthesis Study on Use of Concrete Recycled from Pavement and Building Rubble in the Indiana Highway System*. Joint Highway Research Program, Purdue University, West Lafayette, IN, 1992.
7. American Concrete Pavement Association. *Recycling Concrete Pavements*. American Concrete Pavement Association, Skokie, IL, 2009.
8. AASHTO. *Standard Specification for Reclaimed Concrete Aggregate for Use as Coarse Aggregate in Hydraulic Cement Concrete*. AASHTO Provisional Standard MP 16-13. AASHTO, Washington, D.C., 2015.
9. Donalson, J., R. Curtis, and F. T. Najafi. Sustainable Assessment of Recycled Concrete Aggregate (RCA) Used in Highway Construction. Presented at 90th Annual Meeting of the Transportation Research Board, Washington, D.C., 2011.
10. Ram, P., T. Van Dam, J. Meijer, and K. Smith. *Sustainable Recycled Materials for Concrete Pavements*. Michigan Department of Transportation, Ann Arbor, MI 2011.
11. Horvath, A. *A Life-Cycle Analysis Model and Decision-Support Tool for Selecting Recycled versus Virgin Materials for Highway Applications*. Final Report for RMRC Research Project No. 23. University of California, Berkeley, CA, 2004.
12. Nathman, R. K. *PaLATE User Guide, Example Exercise, and Contextual Discussion*. Master's thesis. University of Delaware, Newark, DE, 2008.
13. Verian, K. P., N. M. Whiting, J. Olek, J. Jain, and M. B. Snyder. *Using Recycled Concrete as Aggregate in Concrete Pavements to Reduce Materials Cost*. Report No. FHWA/IN/JTRP-2013/18. Joint Transportation Research Program, Indiana Department of Transportation and Purdue University, West Lafayette, IN, 2013.
14. Hu, J., D. Fowler, M. S. Siddiqui, and D. Whitney. *Feasibility Study of Two-Lift Concrete Paving*. Report No. FHWA/TX-14/0-6749-1. Texas Department of Transportation, Austin, TX, 2014.
15. Reza, F., and W. J. Wilde. *Evaluation of Recycled Aggregates Test Section Performance*. Report No. MN/RC 2017-06. Minnesota Department of Transportation, Saint Paul, MN, 2017.
16. Singh, S. P., and S. Arora. Evaluating the Flexural Fatigue Performance of Concrete Made with Coarse Recycled Concrete Aggregates. *Proc., 9th International Concrete Conference*, University of Dundee, Dundee, Scotland, 2016.
17. Akkari, A., and B. Izevbekhai. *Composite Pavements at MnRoad: Cells 70, 71 and 72 Construction Report and Early Performance Evaluation*. Minnesota Department of Transportation, Saint Paul, MN, 2011.
18. Izevbekhai, B. Innovative Concrete Mixes in MnROAD Phase 2 Initiatives. *Proc., 2015 Minnesota Pavement Conference*, Saint Paul, MN, 2015.
19. MNDOT. *Average Bid Prices 2014*. https://www.dot.state.mn.us/bidlet/avgPrice/142014.pdf.

The Standing Committee on Concrete Materials and Placement Techniques (AFN40) peer-reviewed this paper (18-05474).

The views expressed here are those of the authors and do not necessarily reflect the view of MnDOT.

Sage Publications USA
sagepub.com

ISBN 978-1-5108-7831-0

CURRAN ASSOCIATES INC.
proceedings.com